T0231767

Practical Multi-Projector Display Design

Practical Multi-Projector Display Design

Aditi Majumder
Michael S. Brown

CRC Press
Taylor & Francis Group
Boca Raton London New York

CRC Press is an imprint of the
Taylor & Francis Group, an **informa** business

AN A K PETERS BOOK

Editorial, Sales, and Customer Service Office

CRC Press
Taylor & Francis Group
6000 Broken Sound Parkway NW, Suite 300
Boca Raton, FL 33487-2742

First issued in hardback 2019

© 2007 by Taylor & Francis Group, LLC
CRC Press is an imprint of Taylor & Francis Group, an Informa business

No claim to original U.S. Government works

ISBN-13: 978-1-56881-310-3 (hbk)

This book contains information obtained from authentic and highly regarded sources. Reasonable efforts have been made to publish reliable data and information, but the author and publisher cannot assume responsibility for the validity of all materials or the consequences of their use. The authors and publishers have attempted to trace the copyright holders of all material reproduced in this publication and apologize to copyright holders if permission to publish in this form has not been obtained. If any copyright material has not been acknowledged please write and let us know so we may rectify in any future reprint.

Except as permitted under U.S. Copyright Law, no part of this book may be reprinted, reproduced, transmitted, or utilized in any form by any electronic, mechanical, or other means, now known or hereafter invented, including photocopying, microfilming, and recording, or in any information storage or retrieval system, without written permission from the publishers.

For permission to photocopy or use material electronically from this work, please access www.copyright.com (http://www.copyright.com/) or contact the Copyright Clearance Center, Inc. (CCC), 222 Rosewood Drive, Danvers, MA 01923, 978-750-8400. CCC is a not-for-profit organization that provides licenses and registration for a variety of users. For organizations that have been granted a photocopy license by the CCC, a separate system of payment has been arranged.

Trademark Notice: Product or corporate names may be trademarks or registered trademarks, and are used only for identification and explanation without intent to infringe.

Visit the Taylor & Francis Web site at
http://www.taylorandfrancis.com

and the CRC Press Web site at
http://www.crcpress.com

Library of Congress Cataloging-in-Publication Data

2007015705

Majumder, Aditi.
    Practical multi-projector display design / Aditi Majumder, Michael S. Brown.
        p. cm.
    Includes bibliographical references and index.
    ISBN 978-1-56881-310-3 (alk. paper)
    1. Image processing–Digital techniques. I. Brown, Michael S. II. title.
    TA1637.M325 2007
    621.36'7--dc22

We acknowledge Kodak for providing us with the picture projected in the image on the front cover. We also acknowledge Argonne National Laboratory for letting us use their 15-projector ActiveMural to display the picture. The image is Figure 4.4 in this book.

Contents

Foreword

YOU HOLD IN YOUR HANDS A VALUABLE BOOK, one that can guide you into an exciting new area, a field that may, dare I say it, change the world. The title is modest enough, *Practical Multi-Projector Display Design*, but it may liberate our thinking about displays. We now think of a computer display as a page-size rectangular area—a small page like on a PDA to a very large page like a 55-inch HDTV, but a page all the same. This book may liberate us to thinking about a display as a wall-sized grid of projectors—a mural, not a page. Later, we may realize that the multiple projectors can create detailed imagery all around us, from 360-degree surround to projecting images on every surface in an office, from the file cabinet to the trash can.

Why might there be a technological revolution in store? The primary reason is the gross imbalance in today's computer technology between displays and all the other key components of personal computing. Consider the major components: processing, main memory, secondary memory (hard drives), displays, and network communications. In the past 25 years, all of them except for displays have exploded in capability: processors from 1 MHz to multiple GHz; memory from under 1 MB to multiple GB; hard drives from 100 MB to hundreds of GB; and network speeds from MHz to GHz. The poor display has increased from VGA to XSGA+, from 0.5 Mpixels to 1.5 Mpixels—a factor of just 3!

Until recently, we could excuse this imbalance based on the sheer expense of displays. Multiple Mpixel displays couldn't be manufactured, and projectors were prohibitively expensive. Until a few years ago, a 1 Mpixel projector might have cost $10,000. A 0.5 Mpixel projector might be affordable for a meeting room, but it was too expensive for an individual office.

A skeptic might argue that displays are fine the way they are today; they're improving steadily, and anyway, there's no law of the universe that demands that display capabilities increase in proportion with the rest of the computing environment; a user can't look at more than a page or two at a time in any case.

Many of us would disagree, of course, arguing that opportunities abound for greater use of displays. Let's consider a range of everyday work situations for a start: an individual office, a group meeting room, a classroom, and a large professional conference.

A modern individual office comes equipped with a desktop PC with a flat-panel display, typically with a 17-inch to 21-inch diagonal. Unless the office inhabitant is engaged in visually intensive work such as graphic design or video production, there is rarely more than one of these displays in the office. Even for non-specialists, many everyday tasks require frequent manipulation of multiple windows and contexts. Consider formulating a reply to a simple email request to evaluate several suggested dates for a meeting a few months hence. The user probably needs to consult a personal calendar, scrolling around in it to check each of the potential dates, then needs to keep the email reply window open to be able to insert a comment by each potential date. In fact, in some ways, this is a more awkward procedure than it used to be when one's annual date book was a physical volume and one's annual calendar was mounted on the wall and marked with a pen. In the future, with projectors displaying pixels everywhere in the office simultaneously, one might again benefit from an annual calendar (marked up with the latest information) mounted/projected on the wall.

For another example, consider a meeting room. In the "old days," there might be easels with large sheets of paper around, so that in a brainstorming session, one could tear off a sheet of paper one wished to keep and mount it somewhere on the wall. After some time, there might be multiple sheets of paper that could be readily consulted. In today's meeting room with a single projector, only a limited amount of information can be displayed at any one time.

Similar limitations exist in the classroom; only a single projected image is available in most classrooms. Old lecture halls, with multiple blackboards, often wrapping around the room, enabled the instructor to write and not erase during the entire lecture, allowing the students to view the material during the entire lecture. Today's version, where the instructor hands out the slide notes, allows the students to flip back and forth but not to have the entire class and the instructor all focus on any material simultaneously. Extemporaneously relating material between two different slides requires the instructor and the students to flip back and forth between the two arbitrary slides. If only projectors would be able to display pixels everywhere simultaneously, the classroom setting could have the best of both worlds.

Professional meetings share some of these same display limitations. Consider the perennial advice about not squeezing too much information onto a single slide ("don't put more than six lines per slide, and no more than six words per line"). This is about the amount of information you can put on the smallest of Post-it Notes (1.5" × 2"). At a normal reading distance, this tiny Post-it Note appears to be about the same size as the projected image of a presentation slide seen from the back of the room at a large professional meeting. How sad that we can't do any better than this.

We might be able to do better in the near future, and the topics in this book provide some of the opportunities. The projectors, which until recently were prohibitively expensive, have become sufficiently affordable that many professionals can afford more than one of them. And as their prices continue to plummet, we can consider having dozens of them in a meeting room or other multi-user facility, perhaps eventually dozens in an individual office. What can we imagine doing with that many projectors?

Most basically, we can combine multiple projectors to form a single large, bright, high-resolution, wall-sized display surface. The techniques for doing this, correcting the geometric distortions, and calculating the appropriate blending functions have been worked out reasonably well, and they are clearly presented in this book. Also important are techniques for rapidly setting up and changing the arrangement of such a wall-sized array of projectors. These techniques will enable major improvement in displays in a multitude of situations—in offices, in meeting rooms, in classrooms, and at conference sites.

In addition, much work remains to be done, and there is plenty of room for innovation: projection onto arbitrary surfaces, adaptation to a changing projection environment, and efficient distributed rendering, to name a few. Much of the work that remains is not just in the area of computer graphics but in the domain of distributed computing and operating systems—the challenge, for example, of controlling a multitude of projectors from a single laptop. While pioneering systems such as Chromium demonstrated how to get started, much work remains. Consider, for example, enabling multi-projector capability not just for graphics applications but for all applications run on the laptop. The support will have to be integrated with the laptop's operating system—a daunting challenge.

Remember, also, that we don't want to be restricted to displaying onto a single wall, but everywhere, about a 360-degree field of surround. The projectors might be mounted like track lights, placed and directed casually to wherever the user pleases. The projectors would light most important surfaces in the room, just like conventional lights in a room currently light most important surfaces. We might then have sufficient pixels and surface area to display not just the "essential digital information" such as email or PowerPoint presentations, but everyday prosaic matters such as an annual calendar, with important dates marked, travel dates circled. Having cameras in the projector units can enable users to mark up the digital calendar with styli or even their fingers. We could then integrate these projector units with conventional LCDs; we could project onto real desktops and integrate with real, physical documents and other objects, for example, cutting and pasting between "real" and digital media. Pioneering work in each of these areas gives us a glimpse of the future, but many more individuals and groups will have to be inspired and engaged if the dreams are to be realized.

We can engage with other emerging research communities, such as those engaged in ubiquitous computing. If most surfaces in an office are to be covered by projectors, most surfaces will have to be observable by cameras in order to calibrate and blend the projected images. Those cameras could be recording and processing the images continuously for future data extraction. Assuming privacy concerns are properly addressed, consider how useful such data would be for most of us. We could retrieve information about past events in our office, recalling information often difficult to find now—"What was that article I was reading last year about...?" or "What was the name of the visitor last month? I remember he gave me a business card."

If all of these projectors, cameras, and processors seem way too expensive, we should consider how expensive simple personal computers must have seemed when they were first being implemented, about 25 years ago. In 1982, computers that were truly personal, such as the Apple II, were genuinely toys. Personal computers that were powerful enough for most professional use, "workstations," were $50,000 or more "per seat"—"seat" because such computers were desktop machines that one sat next to. Today, as we all know, increasing numbers of high schools issue a tablet to each student at the beginning of the school year the same way they issue textbooks. The dramatic improvement in utility is not simply due to the drop in cost but to the ubiquity of the machines. Students uses their tablets not just for "professional" uses such as document preparation, spreadsheets, and computations like the professionals of a generation ago used their workstations; they also use the tablets for all kinds of other activities from morning to night—from personal communications and playing music to volunteer work at a political campaign. It's now a reasonable assumption in many high schools (and, of course, in colleges) that each student will have a laptop.

With some luck, the affordability and ubiquity of multi-projector displays will follow a similar path to that of the personal computer. We should keep in mind, however, that even with the personal computer, the path was much longer than even 25 years. In 1968, Alan Kay described the vision of a Dynabook, a personal computing tablet simple enough for schoolchildren to use, powerful enough to serve as one's principal computer, and useful and affordable enough for schools to issue to each student.

It's tempting to believe that a similar glorious future lies ahead for displays. We're confident that a meeting room of the future won't just remain with a single ceiling-mounted projector, that the office of the future won't just have one or two flat-panel displays. But what the places will look like in the future, we can't be sure. It may be sobering, but prudent, to recall that some futurists predicted in the 1950s and even earlier that soon there would be a "helicopter in every garage."

Essential to the excitement is (1) not knowing how the future will unfold—like predictions of the Dynabook for each schoolchild or the helicopter in every garage—and (2) the realization that we each have the potential to influence what will happen. As Alan Kay so famously put it, "The best way to predict the future is to invent it."

With this book in hand, you have the tools to start. Enjoy.

Henry Fuchs

The University of North Carolina
at Chapel Hill

1

Introduction

I N THE EARLY 1980s, Bill Joy, the co-founder of Sun Microsystems, envisioned a "3M" computer with a megahertz processor, megabytes of memory, and a display of several megapixels. In just two decades, increases in processing power and memory have not only come to fruition, they seem humble in comparison with today's gigahertz processors and gigabyte memory. The megapixel display, however, has only barely met its milestone, with most current computer displays being approximately one megapixel (1000×1000) in resolution.

It is interesting that although we have seemed to accept that our daily computing experience is limited to a 19-inch "through-a-window" display paradigm, other aspects of computing continue to widen the gap of this computing-to-display asymmetry. The capture of very high-resolution data is both affordable and accessible today. Inexpensive scanners and digital cameras have enabled virtually anyone to produce rich and high-resolution media. In the scientific and engineering realms, sophisticated sensors are capturing very large amounts of multimodal high-resolution data for scientific-visualization applications. Nuclear and astronomy databases, on the order of petabytes (10^{15} bytes) in size, are generated at a monthly or even weekly basis at different US National Laboratories such as Sandia, Argonne, and Lawrence Livermore. Outside the sciences and engineering, our libraries, museums, and governmental archives are scanning their precious materials at very high resolution and making this media available to the public.

Yet, current desktop monitors fall short, both in terms of scale and resolution, even for relatively low-end applications like displaying high-resolution digital images. Needless to say, high-resolution displays with at

1

Figure 1.1. A large-area display at Argonne National Laboratory. (From [58]. © 2005 ACM, Inc. Included here by permission.)

least a couple of orders of magnitude more pixel density than the current monitors are critical for high-end applications such as scientific visualization. In addition, large field of view and life-size imagery are essential to mimic realism in applications such as tele-collaboration, tele-medicine, and virtual-reality environments for training, simulation, and entertainment. Simply put, for many applications, better, large-scale displays are needed.

Figure 1.1 shows a large-scale display at the Argonne National Laboratory. Compared to a 19-inch monitor or flat-panel display with a resolution of 100 pixels per inch, displays such as this one cover a large area of about 15 × 10 feet in size and provide resolutions in the range of 200–300 pixels per inch, providing users with a few billion pixels. Such a large-scale display offers vivid, life-size high-resolution imagery more proportional to the datasets at hand, and the display's scale facilities group interaction and invites data exploration.

While the current mindset is that such displays are luxury items reserved for visualization in research and engineering labs, we find it curious that such displays are not more prevalent, especially given the richness and resolution of media in our daily computing experience. Why does our local library not have a high-resolution display, or our children's primary and secondary schools? Why do companies not have a large-scale display for teleconferencing with remote meetings? Why are our homes and offices still limited to small and low-resolution displays?

With this book, we hope to show that recent advances in large-scale display design are making it much more practical and affordable to set up, maintain, and operate large-scale displays. While it will take some time for the costs involved in a large-scale display to reach those of an off-the-shelf desktop display, the traditional costs and difficulties associated with deploying and maintaining large-scale displays are rapidly being reduced, and we hope that in the near future, our dream computer's display may catch up to the other components.

1.1 Life-Size High-Resolution Displays

Providing both life-size imagery and the desired pixel density in an effective large-scale display is no minor requirement. The size of any digital data, be it images, tables, or multidimensional entities out of scientific applications, has increased by orders of magnitude in the last decade, and the laptop or desktop monitors today fall short of providing the screen real estate and resolution required to study and show data in an effective and efficient manner.

There is no *single* display existing today that can meet such taxing demands of both scale and resolution simultaneously. Commercially available large-screen TVs and flat-panel displays 60 inches or so in diagonal are available but offer fixed resolutions of only about 35 pixels per inch. Even the 56-inch LCD Quad HDTV to be released in the summer of 2007 has a resolution of about 80 pixels per inch. High-end flat-panel monitors, such as IBM's T221 LCD monitor, offer very impressive resolution on the order of 200 pixels per inch but have a size of about 22 inches diagonal.

Given the limitations of single display devices, it is not surprising then that current large-scale high-resolution displays are made up of more than one display device. Such design is similar to the path taken in supercomputer design, which moved away from monolithic centralized architectures to the parallelization of hundreds of smaller and cheaper components. Current large-scale displays are usually made by tiling multiple LCD panels or projectors in a two-dimensional rectangular array to create one single large display. LCD panels can be engineered to be mounted on a single substrate to create a display that does not require much maintenance. The pixels can be aligned perfectly across boundaries during this manual construction. However, the bezels bordering each panel make it difficult to generate a seamless display using multiple such panels.

High-end applications, especially for training and simulation, desire or demand a seamless large high-resolution display. Even for visualization, tele-medicine, and tele-collaboration applications, scientists prefer to work with a seamless display where they do not need to consciously move the data around to avoid the bezels. As a result, projection-based displays are often preferred over LCD panels. In addition, projection-based displays are flexible for projection onto arbitrary surfaces. Another added benefit is that projection-based displays are scalable, both in size and resolution. The size of the display can be easily increased by adding more projectors. More importantly, the pixel density can be easily increased by changing the display field of view of the projector. Multi-projector displays can be reconfigured to provide a different aspect ratio by changing the way the projectors are arranged in a two-dimensional array. Finally, if required, such displays can be much easier to dismantle in one location and then set up in an entirely different location.

1.2   Building Large Tiled Displays

The easiest way to build a multi-projector display is to arrange projectors in a two-dimensional array on inexpensive racks or shelves. Figure 1.2 shows such an infrastructure for a nine-projector display at the University of California, Irvine. The projectors are arranged in a 3×3 array projecting on a screen 10×8 feet in size. This results in a relatively low resolution of 30 pixels per inch.

Projector displays can be arranged in two different ways, as shown in Figure 1.3. The projectors can be arranged carefully to abut each other. In this configuration, a slight mechanical movement in the projector position leads to a doubly bright seam at the projector boundary. To avoid this mechanical rigidity, the projectors can be arranged so that they overlap at their boundaries. However, this leads to the introduction of high brightness overlap regions, which then need to be appropriately handled.

It is obvious that rendering the data on a display made of more than one unit is different and more complicated than rendering on a desktop. Figure 1.4 shows a specific pipeline of a large high-resolution data set being rendered on such a multi-projector display. This particular pipeline shows a centralized architecture where a central client process divides the large high-resolution data that is to be displayed on the tiled display into several smaller parts. These smaller images are then shipped to server PCs via the network. Each of these PCs drives one projector, which projects or renders these images on the display screen to be seen by a viewer.

Figure 1.2. A multi-projector display at University of California, Irvine. The display is driven by a nine-PC cluster, each PC driving one projector. Left: the set-up; right: the display from the front as seen by the user. (© 2006 IEEE. Reprinted, with permission, from [7].)

Figure 1.3. Left: abutting projectors. Right: overlapping projectors. (© 2000 IEEE. Reprinted, with permission, from [60].)

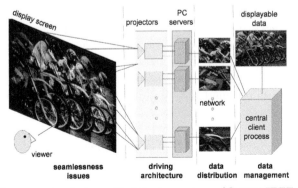

Figure 1.4. Components of a large tiled display system. (© 2006 IEEE. Reprinted, with permission, from [7].)

Let us take a moment to examine this centralized architecture. Note that this rendering pipeline spans various domains of computer science such as data management, networking, graphics, and even computer vision. The display has several different parts, each of which is a challenge by itself. Generating the imagery to be displayed from the raw data and handling and processing the hundreds of millions of pixels thus generated necessitate efficient *data management* techniques at the server end. Shipping this large amount of data efficiently to the different PC clients demands sophisticated *data distribution*. *Efficient architecture* is essential for achieving real-time interactive performance, especially when used for virtual-reality environments. Finally, the display needs to be perfectly "seamless" and undistorted, in terms of geometry and color. This book provides you with all of the fundamentals required to build and set up your own multi-projector display.

At this point, let us take a moment to condier the issue of "seams" in multi-projector displays. Seams of any form in a multi-projector display break the illusion that the user is seeing one large display and not an assembly of multiple display units. Seamlessness is especially critical for applications like training, simulation, and entertainment. Even for less demanding applications like visualization and collaboration, noticeable color variation and geometry breaks or variation act as a distraction causing a hindrance to perform the task at hand. Interestingly, the seams caused by the bezels of LCD panels are perceptually easier to accept. The bezels cause a "French door" effect, and the human visual system can filter out the occluded parts of the scene easily due to years of experience in handling such scenes. In fact, the resulting discontinuity can at times aid this process by hiding small geometric and color mismatches. However, color blotches or geometric breaks in a physically continuous display, such as a projection-based display, is not something we experience regularly and therefore can cause distraction. Thus, geometric and photometric issues for projection-based displays need to be handled effectively to create the illusion of a seamless display. These seams or variations in a multi-projector display arise from two fundamental problems: *geometric misalignment* and *color variation*.

Geometric misalignment. The final overall image of a tiled display is (by definition) formed from multiple individual display devices. These individual devices may not be aligned perfectly at their boundaries, or in the case of arbitrary displays, it can be impossible to achieve alignment. Such misalignments result in visible breaks in the image content across the projector

Figure 1.5. Left: geometric misalignment across projector boundaries. Note the area marked by the yellow square where the fender is broken across projector boundaries. Right: color variation even after the projectors in the display are geometrically aligned with respect to each other. (From [58]. © 2005 ACM, Inc. Included here by permission. Displays courtesy of Argonne National Laboratory.)

boundaries. This is illustrated in Figure 1.5 (left). Note that the fender of the bike appears broken at the boundary between two projectors. This geometric misalignment problem is further complicated by nonlinear radial distortion of the projectors, which tends to bend straight lines into curved lines, making this misalignment easily visible.

Color variation. The color across a multi-projector display can vary significantly (see Figure 1.3). Every pixel in these displays is driven with the same input value, but the corresponding final colors on the display surface are not the same. This illustrates the general problem of color variation in multi-projector displays. Even in the presence of perfect geometric alignment, the color variation is sufficient to break the illusion of a single seamless display (see Figure 1.5 (right). This problem can be caused by device-dependent conditions like *intra-projector color variation* (color variation within a single projector due to distance attenuation of light) and *inter-projector color variation* (color variation across different projectors due to mechanical imprecision in providing identical filters or difference in bulb age) or by device-independent conditions such as a non-Lambertian display surface, overlaps between projectors, and interreflections.

As a result, one of the primary challenges in building an effective multi-projector display is to make the composited imagery from the multiple projectors appear *perceptually seamless*, i.e., as if they were being projected from a single display device. This involves correcting for geometric misalignment and color variation within and across the different projectors to create a final image that appears both geometrically and photometrically seamless to the observer. Determining the necessary per-projector corrections to generate a seamless display from a given setup is commonly referred to as "calibration." Calibration involves two aspects: *geometric alignment* and *color seamlessness*. Geometric alignment deals with geometric continuity of the entire display, e.g., a straight line across a display made from multiple projectors should remain straight. Color seamlessness deals with the color continuity of the display, e.g., the brightness of the projected imagery should not vary visibly within the display.

1.3   State of the Art

Multi-projector displays have been around for almost three decades now, the first one being a three-projector flight simulation system introduced by Evans and Sutherland in the late 1970s. However, until very recently, such displays were restricted to high-tech commercial environments such as flight simulation and entertainment. This was primarily because of the cost and rigidity of the associated infrastructure. Expensive monolithic rendering engines (like SGI InfiniteReality, costing about $1+ million) running application-specific software were used to interface seamlessly with the desired application. Further, geometric and photometric calibration were achieved through expensive mechanical and electronic alignment. Such alignment procedures often required a specialized display infrastructure and a great deal of personnel resources, both to set up and maintain. For example, specialized, expensive light projectors (costing around $100,000 to $150,000) with sophisticated optical elements, optical screens with attached Fresnel lenses ($25,000 for a screen that is 8 feet × 8 feet) were used to achieve view-independent color seamlessness of the imagery projected from multiple display devices. In some cases, custom-made projector mounts (costing $2000 to $5000) with six degrees of freedom (DOF) were used to manually move heavy projectors around and align the images geometrically across the projector boundaries. To top it all, an expensive high-tech maintenance crew was required to install and maintain the display at all times. Thus, having such displays was generally a multimillion-dollar investment with high recurring cost. Further, this prohibitive cost of projectors and driving engines limited the number of projectors to a handful making manual geometric alignment using six-DOF projector mounts and color balancing using expensive optics like Fresnel lenses feasible.

Today, revolutionary advances in projection technology have made high-quality projectors light, portable, and affordable commodity products (costing \$1500 to \$2000). Further, advances in PC graphics hardware enable an ATI or NVIDIA graphics accelerator card to offer performance similar to that of giant monolithic rendering engines of yesteryear at a cost that is many orders of magnitude lower. The driving architecture paradigm has shifted from a single monolithic rendering engine to a multiple-PC cluuster that is not only more affordable but also more accessible and mobile. As a result, today high-resolution displays can be built with significantly cheaper components. Such scalable displays are now being built in-house in academic environments such as universities and national laboratories. Sandia National Laboratory, Lawrence Livermore National Laboratory and Argonne National Laboratory are using such multi-projector displays for visualizing very large scientific data, on the order of petabytes (10^{15} bytes) and higher, and also for holding meetings between collaborators located all around the country via collaborative environments like Access Grid [21, 47, 37]. Such displays are also being used to create high-quality virtual-reality (VR) environments that simulate sophisticated training environments for pilots and military [64]. Such VR environments are also used for entertainment purposes, for example, by Disney. Fraunhofer Institute of Germany, located only a few miles away from the Mercedes manufacturing station at Stuttgart, has at least six such displays, all of which are used to visualize large data sets generated during the design of automobiles or for virtual auto crash tests. Similar displays are being investigated at Princeton University, the University of North Carolina at Chapel Hill, Stanford University, the University of Kentucky, University of California, Irvine, and the National Center for Supercomputing Applications (NCSA) at the University of Illinois at Urbana Champaign and others [53, 40, 103, 79, 12, 87]. Multi-projector displays are also becoming an integral part of user-interface design in ubiquitous computing environments [26, 44, 93, 79, 76, 98, 11, 15, 49, 86, 1, 74, 89, 98, 75]. Steerable projectors realized by mounting them on software-interfaced pan-tilt units can provide interactive user interfaces on different objects in a ubiquitous computing environment.

Today's multi-projector displays thus span a large spectrum in size starting from small ones (4–6 projectors), to medium ones (15–20 projectors), to very large ones (20–60 projectors). When dealing with such a large number of projectors, manual geometric and color calibration become both unscalable and infeasible. Most sophisticated mounting hardware does not have the capability or the precision to correct nonlinear distortions such as projector radial distortion and intensity nonlinearities. Thus, calibrating even a four-projector system manually can be severely time consuming.

Recently, several *camera-based automated calibration techniques* have been developed that use one or more cameras to observe a given display set-up where projectors are only *casually* aligned. Using feedback obtained from a camera observing the display set-up, the necessary adjustments to register the imagery, both in terms of geometry and color, can be automatically computed and applied through software [20, 19, 38, 62, 63, 60, 56, 57, 58, 80, 83, 82, 103, 13, 61, 78, 68]. The key idea in these approaches is to use cameras to provide *closed-loop control*. The geometric misalignments and color imbalances are detected by a camera (or cameras) that monitors the contributions of multiple light projectors using image-processing and computer-vision techniques. From the camera-based observation of the individual projector's contribution to the overall display, geometric- and color-correction functions necessary to enable the generation of a single seamless image across the entire multi-projector display can be determined. Finally, the image from each projector is appropriately pre-distorted by the software in real time to achieve the correction. Thus, projectors can be casually placed, and the resulting inaccuracies in geometries and color can be corrected automatically by the camera-based calibration techniques in minutes, greatly simplifying the deployment of projection-based displays. In comparison with traditional systems relying on precise set-ups, camera-based calibration techniques provide the following advantages.

- More flexility. Large-format displays with camera-based calibration can be deployed in a wide variety of environments, for example, in the corner of a room, or across a column. These surface irregularities can cause distortions that traditional systems may find difficult to work with.

- Easy to set up and maintain. Camera-based calibration techniques can completely automate the set-up of large-format displays. Calibration can now be done in much less time and in a repeatable fashion with little user intervention. This is particularly attractive for temporary set-ups in trade shows or field environments. Further, this allows periodic, automatic recalibration, which makes maintenance of multi-projector displays much easier. Labor-intensive color-balancing and geometric-alignment procedures can be avoided, and automated techniques can be used to calibrate the display in just minutes.

- Reduced costs. Since precise mounting of projectors is not necessary, projectors can be casually placed using commodity support structures (or even as simple as laying the projectors on a shelf). In addition, it is not necessary to hire trained professionals to maintain a precise alignment to keep the display functional. Further, since the color variations can also be compensated for, expensive projectors with high-quality optics (that assure color uniformity) can be replaced by inexpensive commodity ones.

While camera-based calibration techniques require cameras and support hardware to digitize video signals, these costs are amortized by savings from long-term maintenance costs. Overheads like warping and blending at rendering time to correct for various distortions are reduced or eliminated by the recent advances in graphics hardware.

1.4   Focus and Organization

Installing and maintaining multi-projector displays today still requires a technology-friendly user with sufficient expertise in cameras, projectors, calibration methods, operating systems, and user interfaces. This does not foster confidence in a "layman" user to build and use such multi-projector displays in his or her workplace.

The goal of this book is to equip the reader with the knowledge required for practical design, deployment, and maintenance of a seamless multi-projector display. Most of this book assumes the centralized architecture presented in Figure 1.4. Of the different issues mentioned in Figure 1.4, this book focuses on the driving architecture and the seamlessness issues.

Driving architecture deals with the distributed rendering, associated infrastructure, and software (Chapter 5). In this context, we also walk the reader through different practical design options available when deciding on the hardware and the associated accessories for the display (Chapter 2). The seamlessness issue deals with analysis of the cause and nature of the geometric/color seamlessness, followed by methodologies used to achieve a seamless usable display of high quality. The reader is exposed to a large number of available automated camera-based calibration methodologies and their associated costs and benefits (Chapters 3 and 4).

As an interface between the driving architecture and the seamlessness issue, we discuss the implementation of the calibration methodologies in the existing distributed rendering architectures (Chapter 5). Finally, we provide readers interested in knowing the advanced methodologies pushing the frontiers of this technology with some near-future applications of planar multi-projector displays using completely distributed architecture and calibration mechanisms (Chapter 6).

I hear and I forget. I see and I remember. I do and I understand.

This well-known quote, attributed to Confucius, is particularly applicable to the goal of this book. The easiest way to get started building a multi-projector display is to get a camera and a few projectors together and build a small multi-projector display using the techniques discussed in this book. Most readers, especially those reasonably familiar with graphics and image processing, will be surprised at just how easy the approaches described in this book are to implement. A simple projector array can be up and running in a matter of days, especially when existing image-processing libraries such as OpenCV [70] are used for camera capture and to aid in image processing. Moreover, once the algorithms are in place, the calibration techniques are robust and stable and can be used to calibrate the display in a variety of difficult environments and arrangements, yielding very low-cost maintenance. The goal of this book is to see that in the near future, seamless projector-based tiled displays are much more prevalent, so much so that they are found regularly in our museums, libraries, schools, homes, and offices.

2

Elements of Projection-Based Displays

DIGITAL PROJECTION TECHNOLOGY has proliferated in our offices and classrooms for decades and is a familiar technology to most of us. At one time, projectors were considered a relatively bulky and costly hardware, an item that belonged fixed in an office meeting room or lecture hall. Recently, however, projectors have undergone an enormous transformation driven by the consumer market, becoming cheaper, lightweight, even portable, with many models now marketed as "personal projectors." As a result, a variety of projection-based technologies are available for use in designing a tiled display. In this chapter, we attempt to provide an overview of this rapidly evolving technology and shed some light on the various available technologies, their pros and cons, as well as other less common projector attributes that one may consider when purchasing projectors. For a more comprehensive overview on projection technology, we point to an excellent book by Stupp and Brennesholtz [94]. Before we begin this discussion, we first present the most common parameters that are used to evaluate projectors.

2.1 Evaluating Projector Characteristics

There are four parameters (or properties) that are typically used to compare and contrast projectors in terms of image quality: *brightness, contrast, color gamut,* and *resolution.* Two other parameters that are not directly related to image quality but are important when deciding on projectors are *light efficacy* and *throw ratio.* Unfortunately, the metrics used to describe these properties are often confusing, especially when marketed to a consumer. In the following, we try to shed some light on these properties.

2.1.1   Brightness

The brightness of a projector is an estimate of the *maximum amount of light* that the device can generate. Projector vendors usually specify this in terms of *lumens*. This defines the perceived power of light from a solid angle of one steradian. In particular, a single lumen (lm) is defined as one *candela* per solid angle (i.e., lm = cd · sr). The amount of "candle power" in a solid angle is not necessarily the easiest unit to comprehend. To give a mental idea of lumens, a standard 65-watt light bulb generates around 850 lumens, and a standard 100-watt light bulb generates around 1700 lumens. One important point to note here is that projector vendors often take advantage of the direction dependency of lumens to describe the light measured from the *center* of the projector in a perpendicular direction; that is, the brightness is not equal across the image. Most projectors exhibit some degree of *vignetting*, i.e., the darkening of corners of an image. This can be caused by optics (the lens itself), the sensor (many sensors are less sensitive to light that hits the sensor at an angle), or from other causes like a filter or screen or lens hood that shades the corners of an image. In the presence of such vignetting, the lumens measured at the center of the screen at a perpendicular direction is maximum. So, specifying the lumens in this fashion is misleading since it does not provide any estimate of how adverse the vignetting is. A projector can have a high lumen rating, but a sharp vignetting effect can reduce the overall perceived brightness of the projected imagery.

In an attempt to provide a solution for this, the ANSI lumen standard was established; the lumen ratings at nine points within the projector's field of view are averaged to generate the ANSI lumen specification. These nine points are such that they sample the projector's field of view uniformly. Currently, almost all projector vendors specify brightness in ANSI lumens.

2.1.2   Contrast

The contrast or *dynamic range* of a display is defined as the *ratio of the maximum to minimum brightness* produced by the device. This is usually specified by the projector vendors as a ratio, e.g., 800 : 1. One may question the minimum brightness in this context. Shouldn't it be zero when projecting black? So, shouldn't the contrast be infinity? Unfortunately, as we will discuss shortly, most projectors project some light even when they are projecting black. This is called the *black offset*. The black offset sets the minimum brightness to a nonzero value leading to a finite value for the contrast.

2.1.3   Color Gamut

Color is a three-dimensional quantity defined by one-dimensional brightness and two-dimensional *chrominance*. Chrominance is defined by the *hue* and *saturation* of a color. Saturation can be thought of as the amount of white in any color. If there is no white, the color is saturated. Saturation decreases with the increase in the amount of white. For example, pink is an unsaturated version of red. A "baby pink" is much more unsaturated than a regular pink. More formal discussion of color space is available in Chapter 4 and Appendix A, but an informal understanding will suffice for this chapter.

Together, the contrast and brightness parameters provide an estimate of the range of brightness the projector can produce. Similarly, the range of chrominance that a projector can produce can be defined by a triangle called the *color gamut* of the projector. The three vertices of the triangle correspond to the chrominance of the three primary colors. The more saturated the primary colors, the larger the color gamut. A larger color gamut indicates a vibrant and rich image color quality. Currently, color gamut is rarely specified by projector vendors. It can, however, be measured easily by a radiometer.

2.1.4   Resolution

The term resolution is often taken to mean "the amount of detail" and should relate to how much detail the display is able to provide. Usually, this can be adequately defined by the pixel density or the number of pixels per unit length/area on a projection surface. For many devices, however, there exists no fixed relationship with the particular physical object to calculate this measurement in terms of pixels per unit length. As a result, resolution is typically used to specify the *total number of pixels* the device is capable of producing, especially for devices such as digital cameras. For example, it is common to say that a digital camera has a resolution of 2000×3000 pixels. Projectors also fall into this category because they do not have an integrated screen—a projector positioned closer to the display screen yields greater pixel density on the display surface but outputs the same number of pixels. Thus, projector vendors use resolution to mean the total number of output pixels.

In this book, we are also guilty of using the term resolution to refer to both pixel density and the total number of output pixels. However, we will distinguish these two definitions by the context in which they are used and by the unit—pixel density has units of pixels per inch on the display surface, while the device output capability has units of just pixels.

2.1.5   Light Efficacy

In any light-producing device, the total amount of light produced by the emitter may not be used for display. Some of the light is inevitably dissipated, leaked, or absorbed within the system. Light efficacy is a measure of how much light is actually used for display. For example, light efficacy of 80% indicates that only 80% of the light produced by the bulb manifests itself in the brightness of the projected imagery.

2.1.6   Throw Ratio

Projectors are unique because they can display an image on a surface and do not need to have an integrated display surface. The size of the image that is produced is closely related to the focal length of the projection lens. In particular, this measurement guides how far the projector needs to be from the screen to produce an image of a certain size. The critical information for calculating this distance is the view frustum of the projector given in terms of horizontal and vertical angles. However, exact frustum information is surprisingly difficult to come by, even from individual manufacturers' specifications. Instead, many projector vendors now provide a throw ratio that is the ratio of the distance to the screen and the width of the image on the screen. This throw ratio can also be specified as *projection factor*, *lens factor*, or even simply as *the throw*. Since the width and the height of the projected imagery is usually bound by one of the standard aspect ratios of 16 : 3 or 4 : 1, it is easy to calculate the exact size of the image given the throw ratio and the distance of the projector from the screen. Some vendors even provide a Web-based calculator to calculate this. A short throw ratio (e.g., 1 : 1) indicates a large image from a relatively small distance from the screen. Such projectors are ideal for rear-projection systems. A long throw ratio allows the projector to be set up at the back of a large auditorium and still produce an image big enough for the whole audience to see.

2.2   Core Projection Technologies

Current projectors are designed using two basic technologies: *light-emitting* technology such as a cathode ray tube (CRT) and *light-valve* technology, which includes liquid crystal display (LCD), digital light processing (DLP) with digital micromirror devices (DMDs), and digital direct drive image light amplifier (D-ILA) with liquid crystal on silicon (LCOS) devices. The CRT technology is light-on-demand technology where the appropriate amount of light is generated for varying input signal strengths. The light-valve technology is a light-attenuating technology where light is continuously generated at peak strength and then blocked out based on the desired output brightness. Though there are only two main projection technologies, the projector architecture can vary greatly, even when using the same technology, based on the different types of optical elements used and the way they are multiplexed to create the red, green, and blue color channels. In this section, we discuss the two core projection technologies (CRT and light valve). The other associated optical elements are discussed in Section 2.3. The details of the different projector architectures are discussed in Section 2.4.

2.2.1   Cathode Ray Tube (CRT) Technology

Devices based on light-emitting technology use an emissive image source, most commonly a cathode ray tube (CRT), illustrated in Figure 2.1. In a CRT, the *cathode* is a heated filament, not unlike the filament in a normal light lamp. The heated filament is inside a glass *tube* that forms a vacuum. The *ray* is a stream of electrons that naturally pour off a heated cathode into the vacuum. Electrons are negatively charged, and the emitted electron stream is focused by a positively charged *anode* into a tight beam and then accelerated by an accelerating anode. This tight beam of electrons

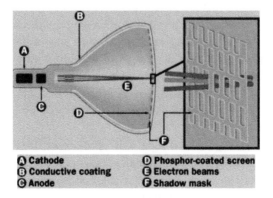

Figure 2.1. A cathode ray tube.

is directed through the vacuum and hits the flat screen coated with phosphorus, called a *faceplate*, at the other end of the tube. The phosphorus glows when struck by the beam to produce various colors. To direct the beam at different locations on the faceplate, the tube is wrapped in coils of wires called *steering coils*. These coils are able to create magnetic fields inside the tube that the electron beam responds to. One set of coils creates a magnetic field that moves the electron beam vertically, while another set moves the beam horizontally. By controlling the voltages in the coils, the electron beam can be positioned at any point on the screen. Generally, this electron beam is swept along the faceplate in a scanline order from top left to bottom right, either in a line-by-line or an interlaced fashion. This electron sweeping is done at a rate fast enough that to an observer, the faceplate appears to glow continually, forming the observed image on the screen. This system that uses an electron beam to light phosphors on the faceplate is called the *light engine*.

2.2.2   Light-Valve Technology

Liquid crystal projectors are based on light-attenuating technology. This consists of a beam separator that splits the white light from a high-powered lamp into red, green, and blue. The split beams then pass through *light-valve panels* that attenuate the amount of light at each pixel differently as per the image input. The number of light-valve panels used can be different, resulting in different projector architectures. The simplest one uses three panels where each of the red, green, and blue beams of light is directed to a different panel. Then, the individual attenuated beams from these panels are recombined using a beam combiner and projected through a projection lens. A schematic illustration of this architecture is presented in Figure 2.2. However, to reduce the cost, a single-panel architecture can be used. Here the red, green, and blue beams are either temporally or spatially multiplexed so that at any point in time, only one of the red, green, or blue beams passes through a pixel of the panel. The beam from the single panel is directly projected via a projection lens. The frequency of multiplexing is made very high so that the colors can be integrated easily by the human visual system.

The core technology behind light-valve projectors, be it three-panel or single-panel, is the light-valve panel made of an array of microdevices. Each of these microdevices corresponds to a pixel and can attenuate the light as per the input at that pixel. The microdevice can be a liquid crystal display (LCD), a digital micromirror device (DMD), or a liquid crystal on silicon

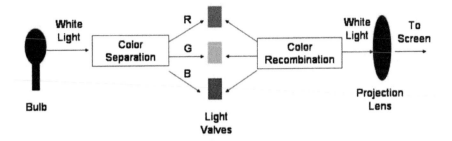

Figure 2.2. Schematic representation of a light valve projection system. (Based on [94]. Copyright © 1999 John Wiley & Sons Limited. Reproduced with permission.)

(LCOS) device. The projection technology using a DMD array was first developed by Texas Instruments, who calls it digital light processing (DLP) technology. Similarly, the projection technology using LCOS devices was first developed by JVC, who calls it the digital direct drive image light amplifier (D-ILA) technology. These are just trademark nomenclatures given by their inventors, but the inherent technology is a DMD and LCOS device, respectively. In this section, we describe the LCD, DMD, and LCOS technologies in details.

Liquid crystal display (LCD). The light valves made of liquid crystal (LC) pixels are addressed by a single transistor per pixel, forming an active matrix device. Voltage applied to the liquid crystal changes its transmissive and dielectric properties, which in turn changes the polarization of light. Usually, the LCs are situated between two alignment layers called the polarizer and analyzer. The LCD technology can be designed in two ways depending on the relative polarizing capabilities of the analyzer and polarizer.

The first is *driven-to-white* mode. In this mode, when no voltage is present, the unpolarized light from the lamp is polarized by the polarizer. This polarized light then passes through the liquid crystal, which changes the polarization direction of the light and is then blocked by the analyzer. When voltage is applied to an individual LC pixel, the liquid crystal's properties change to allow more of the polarized light to maintain its polarization, which decreases the amount of light blocked by the analyzer, thereby increasing the brightness of the projected light. Thus, this mode of having voltage increasing the amount of light projected gives rise to the expression *driven to white*.

Alternatively, there is a *driven-to-black* mode, where no voltage applied to an LC pixel results in the maximum brightness being passed through the analyzer. With applied voltage, the amount of light blocked by the LC increases, and at high voltages, all of the light is blocked. Since this approach produces less light with more voltage, it is said to be *driven to black*.

Digital micromirror device (DMD). A digital light processing (DLP) system consists of an an array of digital micromirror devices (DMDs), where each pixel consists of one such micro-mirror on a torsion-hinge-suspended yoke. The mirror can be deflected by +10 or −10 degrees. In one position, the light falling on the mirror is reflected onto the projection lens. This is called the ON position. In the other, it does not reflect on to the lens. This is called the OFF position. These mirrors can switch between ON and OFF positions at an extremely high temporal frequency. The brightness of a pixel is controlled by modulating the amount of time the mirror is in the ON position, which is often called its *duty cycle*. For example, 50% brightness would require the mirror to have a duty cycle of 0.5 in a given time period. Keeping the mirrors ON continuously until the required duty cycle is achieved and then turning them OFF for the rest of the time period leads to visible flicker. To avoid this flicker, the duty cycle is varied based on the input value by a binary-weighted temporal pulse-width modulation. This is illustrated in Figure 2.3.

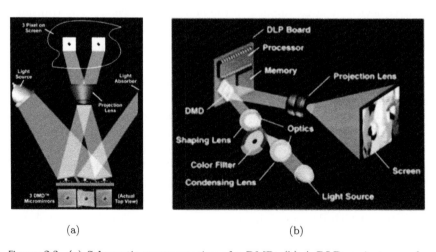

(a) (b)

Figure 2.3. (a) Schematic representation of a DMD. (b) A DLP projector made of DMDs.

Liquid crystal on silicon device (LCOS). LCOS can be thought of as a hybrid between LCD and DLP. LCD uses liquid crystals, one for each pixel, on glass panels. Light passes through these LCD panels on the way to the lens and is modulated by the liquid crystals as it passes. Thus it is a "transmissive" technology. On the other hand, DLP uses tiny mirrors, one for each pixel, to reflect light. DLP modulates the image by tilting the mirrors either into or away from the lens path. It is therefore a "reflective" technology. LCOS combines these two ideas. It is a reflective technology that uses liquid crystals instead of individual mirrors. In LCOS, liquid crystals are applied to a reflective mirror substrate. As the liquid crystals open and close, the light is either reflected from the mirror below or blocked. This modulates the light and creates the image.

2.3 Associated Optical Elements

A projector involves several optical elements in addition to the elements of the core projection technology. These include lenses, filters, mirrors, prisms, integrators, lamps, and screens. These optical elements impact the projected image quality in a nontrivial manner. In this section, we discuss these different optical elements and the artifacts they create for a multi-projector display.

2.3.1 Filters

Filters are used in projectors, often in conjunction with prisms, to split the white light into three primary light paths: red, green, and blue. The important properties of the filters are their spectral response, transmissivity, and reflectivity. Usually, these responses vary with polarization of light and angle of incidence, causing different aberrations that need to be compensated for. There are three types of filters: *absorptive, dichroic*, and *electrically tunable.*

Absorptive filters absorb certain light bands while allowing others to pass through. The absorption of selective wavelengths leads to high thermal stress. This, in turn, leads to lower longevity. Further, absorptive filters have a low light efficacy since they have low transmission capabilities of about 65%. Finally, in practice, it is difficult to design band-pass or short wavelength–pass absorptive filters. Thus, the blue and green primaries often are not saturated leading to smaller color gamut and hence washed-out images. So, absorptive filters are usually avoided in projection design.

Dichroic filters are the most commonly used filters in projection devices, in particular in LCD devices. Unlike absorbtion filters, these filters reflect certain bands of light while allowing others to pass. Using a pair of dichroic filters in tandem, a white light source can be separated into red, green, and blue, where the first filter reflects blue and allows the rest to pass, and the second filter reflects green and allows red to pass. In this manner, both the reflected light and the passed light can be captured, making these filters very efficient. Since most of the light is either transmitted or reflected (almost 96%), this assures a negligible thermal stress.

The biggest advantage of dichroic filters is that they can also be used to recombine the red, green, and blue light paths after passing through the light valves. However, for recombination, the component beams should be distinguished in terms of wavelength or polarization. The way these filters can be used to recombine light is discussed in detail when describing projector architectures in Section 2.4.

The spectral characteristics of dichroic filters tend to vary with angle of incidence. For oblique angles of incidence, especially at more than 20 degrees, the spectral band shows a significant shift to the left. In most projectors, the dichroic filters receive non-telecentric light where the angle of incidence on one end of the filter is different from that at the other end. Non-telecentric light leads to compact design and smaller projectors, as is explained in Section 2.4. But this also leads to a noticeable color shift from one end of the projector to the other due to spectral dependency of the dichroic filters on the angle of incidence. This phenomenon is usually referred to as *coma*. The incidence angle also affects the transmissivity and hence shows a brightness gradient from one side of the projector to the other. As a solution, often a gradient is built into the dichroic filter to compensate for this effect.

Electrically tunable filters have a spectral response that can be modified or tuned electrically, hence the name. When no voltage is applied to these filters, they transmit white light. However, when a voltage is applied, these filters transmit selective wavelengths or polarizations of light. The wavelengths to be transmitted can be changed by changing the composition of the chemical comprising the filter. The light that is not transmitted is reflected back. Electrically tunable filters are the key technology behind single-panel color-field sequential projectors where a color wheel is used to achieve the temporal multiplexing of the red, green, and blue lights. A color wheel is made up of three different electrically tunable filters connected in series. Each filter in this color wheel has a different response, namely red, green, and blue. In each temporal cycle, the voltages going to the three filters are changed to pass either red, green, or blue. The fast switching times of the electrically tuned filters enable high-frequency voltage changes, assuring a high frame rate.

2.3.2   Mirrors, Prisms, and Lenses

Current projection technology is able to produce projectors that are small, lightweight, and portable. This has been one of the key contributing factors in the recent success of projection technology. Mirrors and prisms are the optical elements that make this happen. Reflection from mirrors and total internal reflections in prisms are instrumental in folding the light beam one or more times so that it can be confined in a smaller volume. First, surface mirrors (sometimes referred to as front-surfaced mirrors) are used to achieve this light reflection. Second, surface mirrors produce ghost images, which reduce the effective projector resolution.

The projection-lens system is one of the most important elements of projectors. Two important parameters to evaluate a projector, vignetting and throw ratio, are determined primarily by the type of projection lens being used. Further, since the lens system acts as a nonlinear optical element, it can also introduce nonlinear geometric and color distortions. While zoom lenses with a large range of focal lengths show the greatest distortion, even fixed focal-length lenses can exhibit some types of distortion. The three most common types of lens distortion are *barrel, pincushion,* and *chromatic aberration.*

Barrel and *pincushion* distortion are the most common type of distortion and are caused by nonuniform magnification of the image from the outside of the image (perimeter) to the center. This is illustrated in Fig-

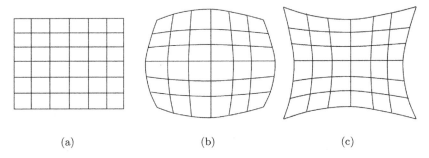

|(a)|(b)|(c)|

Figure 2.4. (a) A checkerboard image with no radial distortion. (b) The same image in the presence of barrel distortion. (c) The same image in the presence of pincushion distortion.

ure 2.4. Barrel distortion refers to magnification that diminishes towards the edges of the image, resulting in an image that looks rounded, like a barrel. Pincushion distortion causes an image to appear pinched or narrowed at the sides.

Chromatic aberration distortion (known as CA distortion) is a result of nonuniform bending of light of varying color (wavelength) as it passes through a lens. Zoom lenses, particularly at their widest and longest focal lengths, exhibit the most severe distortion. This distortion appears most at image corners in high-contrast areas, like branches of a tree silhouetted against a bright sky, and is seen as uneven colors around the details of an image. This is typically called color fringing and is mostly seen in purple colors. Finally, one of the causes of the *vignetting* artifact of darkening of the fringes of an image is the optics of the lens itself.

2.3.3   Integrators

Integrators are a lesser known element of the optical system in light-valve projectors. These are usually used to reduce the vignetting effect by making the brightness of the projected image close to spatially uniform and are usually placed between the lamp and the light valve. They have two primary functions. First, they aim to produce a uniform illumination of the light valve. Second, they convert the circular cross-section of the lamp output beam to the rectangular format and size of the light valve. There are two types of integrators: *lenslet* and *rod*.

Lenslet integrators consist of arrays of very small lenses (lenslets). The lenslet integrator comprises two lenslet arrays following the lamp. The second array is combined with an auxiliary lens. The lenslet array achieves the effect of converting a single source into a spatial array of smaller sources. Thus, it maintains the angle of the illumination cone but redistributes the higher brightness at the center towards the fringes, creating a more spatially uniform light field. However, the number of lenslets used is critical. With too few lenslets, the illumination is not sufficiently uniform, but with too many lenslets, there is no transition region between adjacent lenslets, which creates some step artifacts.

Rod integrators consist of a pair of square plates placed perpendicular to the lamp. The beam from the lamp gets reflected back and forth across these plates before reaching the light valve. If the rod is long enough, the spatial correlation of the beams is lost by the time they reach the end of the rod integrator, creating a more uniform illumination. In fact, with a sufficiently long rod, a perfectly uniform field can be achieved. However, lenslet integrators are often preferred over the rod integrators because a long rod path length can result in a less compact design. In addition, the multiple reflections can result in significant loss in light, reducing the overall light efficacy of the system. Finally, rod integrators can change the polarization of light and therefore can adversely affect the performance of LCD projectors.

2.3.4   Projection Screens and Lamps

The projection screen, while a passive device, is instrumental in redirecting energy in an efficient manner to increase the visual experience. Screens are often described by their *gain*. Gain is a number used to compare a given screen relative to an ideal Lambertain screen. A Lambertain screen reflects (for front projection) or transmits (for rear projection) light equally in all directions. The gain of a screen is defined by the ratio of the light reflected by a screen towards a viewer perpendicular to it to that reflected by a Lambertain screen in the same direction. Thus, a Lambertain screen has a gain of 1.0. It is common to see screens whose gain range from 0.8 to over 2.0. Screens with higher gains direct light in the direction perpendicular to the view rather than distributing it in all direction like a Lambertain screen. This results in a view-dependent image quality, where the image appears dimmer as the angle increases from the perpendicular view. Alternatively, one may use a projection device to project directly onto existing surfaces, such as a white painted wall. Surprisingly, nongloss latex painted walls can provide a reasonable Lambertain effect. High-gloss paint or whiteboards, on the other hand, are very specular and typically will result in view-dependent bright spots on the screen.

Finally, lamps are also a critical component of projection devices. Usually, lamps are accompanied by a reflector that surrounds the lamp and collects the light from all directions and directs it towards the light valve and projection lens. Thus, a reflector is instrumental in increasing the light efficacy of the system.

The properties of a lamp that have a direct effect on projector characteristics are lamp power, lamp efficiency, and spectrum of the light generated. The lamp efficiency is usually measured by the amount of visible light output from the lamp per electrical power input, measured in lumens/watt. Spectral emission lines are one of the most important chromatic artifacts dependent on the light spectrum. Some lamps emit light near yellow and blue, which cannot be blocked by the filters and reduces the saturation of either the green or red and, hence, affects the color gamut adversely. Further, manufacturing errors make it difficult to predict whether the yellow will be included in the green or the red. Thus, different projectors using the same lamp can show different characteristics of the primaries. The same problem near the blue-green region in not nearly as serious because of the considerable overlap in the blue and green spectrums.

Other important lamp properties include the following: arc or filament size—smaller arc size is conducive to compact design; longevity—lamps exhibit reduction in brightness and color shift with age and are considered unusable if they show a 50% reduction in brightness and a major color shift; warm-up time—time required after switching on for the lamp to produce light at the maximum brightness; safety—whether it is safe to use the lamp at different input voltages. Lamp life is an important concern for projector designers. Projector life is likely to exceed 10,000 hours. Currently, a 1000-hour life is considered the minimum required longevity for a projector lamp. The industry target is 10,000 hours so that no lamp change is required during the life of the projector. Currently, there are no lamps that can satisfy this criterion. However, low-power pocket projectors that use LEDs as their bulbs have been recently introduced into the market. These easily achieve this target longevity.

Currently, four types of lamps are used in projectors, including xenon lamps, metal halide lamps, UHP (ultra-high performance) lamps, and tungsten halogen lamps. Metal halide lamps have very high light efficacy (100 lumens/watt) when compared to xenon or tungsten halogen lamps (25 lumens/watt). Metal halide lamps, however, have poor colorimetry due to the presence of spectral emission lines, long warm-up times, large arcs, and major color shift with age. Xenon lamps are better in all of these respects. Their small arc and excellent spectrum make them desirable for applications where total input power is only of secondary importance. Tungsten halogen lamps show a spectrum considerably biased towards the lower wavelengths (bluish) for high temperatures. The significant change in spectrum with temperature (continued use) and relatively low life explain the relatively infrequent usage of these lamps in projectors. The UHP lamp, introduced recently, offers many desirable features. It has a short arc, a relatively high lamp efficiency of 60 lumens/watt, and longevity of around 4000–8000 hours.

Operation of any lamp at a power significantly above its specified power can reduce the lamp life dramatically, sometimes to zero. In the case of some lamps, like xenon lamps, they can also pose a serious safety hazard due to the tendency of the lamp to explode at higher powers.

2.4 Projection Architectures

In this section, we discuss the different CRT architectures, followed by the various light-valve projector architectures.

2.4.1 CRT Projectors

A CRT projector is usually composed of three separate light engines, one for each color channel, each of which is coupled with a projection lens that directs the image from the faceplate to the lens system. The orientation of the three light engines is engineered appropriately so that the images created by the three independent engines converge on the screen to create one crisp image, as illustrated in Figure 2.5(a).

However, this convergence of pixels from different channels is a significant issue, especially in the presence of the geometric distortions caused by the projection lens. This is illustrated in Figure 2.6. As a result, a single-lens architecture is often used, as shown in Figure 2.5(b). Here, the output of the three tubes are combined using dichroic filters and then passed through the projection lens. Different distortions of different channels are avoided, and convergence is much better. Moreover, the convergence need not be adjusted with change in distance from the screen and can be factory-set, instead.

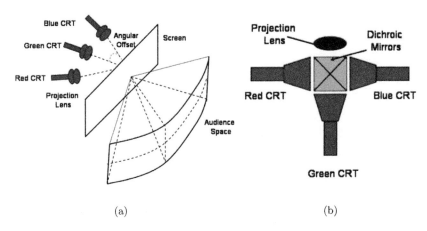

(a)   (b)

Figure 2.5. (a) Schematic of CRT projection system using three lenses. (b) Three CRT single-lens architecture. (Based on [94]. Copyright © 1999 John Wiley & Sons Limited. Reproduced with permission.)

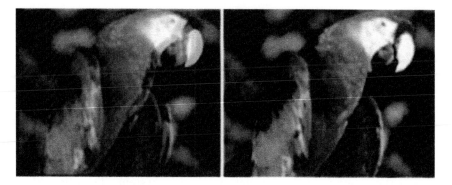

Figure 2.6. The convergence problem of a CRT projector (left) compared to a light-valve projector (right).

2.4.2   Light-Valve Projectors

The architecture of light-valve projectors depends primarily on the number of panels used. Panels are expensive, and using fewer panels results in more affordable devices. In a three-panel architecture, three different panels are used to create the images for three different channels. The white light from the bulb is divided into three parts and directed to appropriate panels, recombined after passing through the panels, and then projected onto the lens. In a less expensive one-panel architecture, a single panel is used, and a color-field sequential system is used to split the light into temporally multiplexed red, green, and blue beams, which are directed to the same panel. The multiplexing is done at a very high frequency so that it is easily integrated by the human visual system.

Three-panel systems. The three-panel system uses three valves, a single lens, and a single lamp. Each panel is dedicated to one color. White light is split into red, green, and blue beams. Each of these beams is directed to a panel. The panel modulates the beam to create the red, green, or blue component of the image. These three images are recombined and then passed through a projection lens. Since, at any instant in time, all the light is used by directing towards one of the panels, this system has a high light efficacy.

The most common architecture for three-panel systems is illustrated in Figure 2.7. In this architecture, dichroic filters are used to split white light to red, green, and blue beams. Further, crossed dichroic filters mounted on the inner diagonal of a prism are used to recombine the split light after

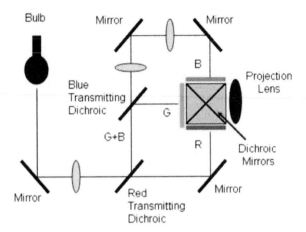

Figure 2.7. Three-panel architecture. (Based on [94]. Copyright © 1999 John Wiley & Sons Limited. Reproduced with permission.)

passing through the panels. The three panels are mounted on three sides of the prism, and the projection lens is mounted on the fourth side. Any three-panel system raises the concern of convergence. The beams coming from the three different panels need to converge perfectly even after passing through the projection-lens system. In this architecture, the panels are mounted on the prism, which makes them shock-resistant. Each panel can be mounted on a six-axis adjustable mount, which can be adjusted in the factory to produce convergence. However, some of the disadvantages of this architecture are as follows. First, the light hitting the dichroic filter is not telecentric, and since these filters are sensitive to angle of incidence, this gives rise to coma. Further, note that the three colors have unequal optical paths, i.e., the path from the lamp to the light valves and from the light valves to the lens are not equal for every channel. Due to these unequal path lengths, additional relay and field lenses (shown in yellow in Figure 2.7) are required in the longer path to get similar illumination.

One-panel systems. As the name implies, one-panel systems use only a single panel. To handle the three color channels, the most common architecture has the red, green, and blue beams *temporally multiplexed* so that only one of them utilizes the panel at any instant. These are called *field-sequential* systems. The temporal multiplexing in this architecture can be achieved by a color wheel rotated by a motor that presents the red, green, and blue filters in succession to the lamp. Thus, at any given time, one of the filtered red, green, or blue beams light the panel. The biggest advantage of these systems is their natural convergence via temporal multiplexing where the images from the three channels are integrated in the human visual system. Field-sequential systems are the most common one-panel systems for DMD projectors. However, an important point to note here is that some of the recent DLP projectors use a clear filter on the color wheel to project the grays and therefore use a four-filter color wheel. With this clear filter, the projector ceases to behave as a normal three-primary system. As a result, the basic assumption that the light projected at any input is the optical superposition of the contributions from the red, green, and blue primaries is no longer valid.

Panels are the most expensive elements of a projector. Hence, one-panel projectors are much less expensive. However, this does come at some disadvantage. First is the cost to be paid in terms of higher data rates (for time-sequential operation). Second, the energy falling on the panel is three times that in three-panel systems. Some polarizers are unable to take this stress, and this extra thermal load on the polarizer may restrict the lamp power used in these projectors. However, the cost and practical feasibility outweighs all of these disadvantages, making this the most popular architecture for commodity projectors.

2.5 Comparison of Different Projection Technologies

Each projection technology has its own advantages and disadvantages in terms of the quality of image generated. When making a choice of hardware for the multi-projector display infrastructure, it is useful to understand the effects of the different projection technologies on image quality. In this section, we discuss the advantages and disadvantages of the different projection technologies.

The brightness of CRT projectors is often limited by the properties of the phosphors. Phosphors do not increase in brightness linearly with the voltage applied to the steering coils. Rather, the brightness gets saturated after the voltage has reached a certain threshold. In addition, CRT projectors suffer from the problem of thermal quenching. At very high temperatures, the emissivity of phosphors decreases with an increase in the beam current. A liquid cooling is coupled with the faceplate to avoid this problem. Together, these phenomena essentially limit the brightness of CRT projectors. Usually, light-valve projectors can be brighter since increasing the power of the lamp can easily increase the brightness. However, projectors with very high-powered lamps also need a cooling system, typically achieved by using a fan.

Inherently, CRT projectors produce light on demand and hence produce no discernable light for black. Thus, as the minimum brightness produced is very close to zero, these projectors usually have very high dynamic range, as high as 3000 : 1 or 4000 : 1. On the other hand, since light-valve projectors use light-blocking technology, producing a "true" black (no light at all) is almost impossible since it means that all of the light that is blocked has to be absorbed within the projector. Thus, they always produce some discernable *black offset*. This black offset reduces the dynamic range of the display significantly despite having a higher brightness than CRT displays. One of the driving goals of further development of light-blocking projection technology is to drive the black offset as low as possible so that contrasts comparable to CRT projectors can be achieved. Today, it is possible to have light valve projectors with 2000 : 1 contrast. However, one important point to remember here is that ambient light itself can cause the contrast to decrease. So, unless one considers a very dark background, very high contrast ratios won't make a difference. For example, PC monitors have a contrast of 300 : 1, and cinematic environments have a contrast of 1000 : 1.

A lack of convergence of the images from the three channels can result in blurry images, reducing the effective resolution of CRT projectors. However, if the convergence is set correctly, being analog devices with discrete pixels, CRT projectors tend to produce very smooth images. LCD projectors transmit light through tiny pixels and project it onto a big screen. This is a "transmissive" technology, and the image can sometimes look pixelated or blocky. This is referred to as "the screen-door effect" because the image looks like you're viewing it through a screen door. If a pixel burns out, it will display as a black or white dot on the screen, and the only way to fix this is to replace the whole chip. The effect is not visible so much in DLP technology, mainly due to its reflective nature where small "bleeding of light" from neighboring pixels offsets this effect. The best trade-off is achieved in the LCOS technology that is a combination of the transmissive LCD and reflective DLP technology. The perfect balance is struck where bleeding is not large enough to sacrifice resolution but just right to remove the screen-door effect. Further, in field-sequential DLP devices, some people see an occasional flashing across the screen called the "rainbow effect." This happens when the eye cannot successfully integrate the temporally multiplexed red, green, and blue channel images.

CRT projectors provide light on demand and hence have high light efficacy with little leakage. However, light-valve devices usually have low light efficacy because light is continuously produced, and the unwanted light is blocked out. This is worse for LCD projectors than for DLP projectors since the LCD panels are not as efficient as DMD arrays.

Portability is also an important factor when deciding on projection technologies. The biggest disadvantage of CRT projectors is that they are large, bulky, and difficult to move between rooms. They often weigh close to several hundred pounds. Light-valve projectors, on the other hand, are compact and very light (usually less than 20 pounds). Thus, they are easily portable.

In terms of cost, CRT projectors are at least two times more expensive than light-valve projectors. Of the light-valve projectors, LCD and DLP projectors are of comparable price, with DLP projectors being a little more expensive. However, DLPs consume less power, primarily because the technology is dependent mostly on mechanical switching of micromirrors and not on changing the conductivity and polarization capabilities of chemicals.

2.6   Multi-Projector Displays

As mentioned in Chapter 1, the goal of a multi-projector display is to produce life-size and high-resolution imagery, as shown in Figure 2.8. Since the resolution of an individual projector is fixed, the most cost-effective approach for doing this in a seamless manner is to use multiple devices together. The aim here is to get the projectors to work together to pro-

(a)   (b)

Figure 2.8. (a) The twenty-dollar bill on a tiled display. (b) Zoomed in to show that we can see details invisible to the naked eye. (Images courtesy of the University of North Carolina at Chapel Hill.)

duce the illusion of a single projector. Our ultimate goal is to design a multi-projector display such that a viewer cannot determine the number of projectors. In subsequent chapters, we will discuss recent techniques to produce this seamlessness in terms of the image geometry and color. In the following sections, we discuss some issues that one should consider when building projector arrays.

2.6.1 Projector-Array Configurations

Multi-projector displays are typically distinguished as either front or rear projection as shown in Figure 2.9. There are advantages and disadvantages to each. For rear-projection systems, an extra space is needed behind the screen, which typically cannot be used for any other purpose. Front-projection systems, on the other hand, typically have the projectors hanging from the ceiling and do not need any extra dedicated space. In this case, however, it is easy to block the light from the projector, creating unwanted shadows on the display as shown in Figure 2.10. Furthermore, users who look back into the projectors can be hit in the eye with uncomfortably high-powered light.

Another consideration for the projection style is with regards to the display screens. Multi-projector displays are typically built to accommodate group interaction and thus several different viewers. Alternatively, a display could be designed for a single, moving user. For such purposes, Lambertian screens are ideal, as they are best suited for multiple view angles. Interestingly, very few rear-projection screens are available that have

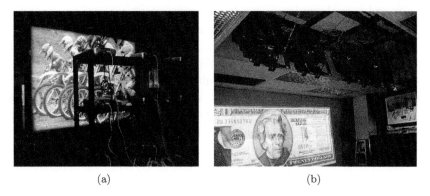

(a)   (b)

Figure 2.9.   (a) Rear-projection system at University of California, Irvine. (© 2006 IEEE. Reprinted, with permission, from [7].) (b) Front-projection system. (Image courtesy of the University of North Carolina at Chapel Hill.)

Figure 2.10. Shadows formed in front-projection systems. (Image courtesy of the University of North Carolina at Chapel Hill.)

screen gain close to 1.0. Not surprisingly, rear-projection set-ups typically show larger brightness variation. At the same time, the diffuse nature of the front-projection screen allows light to spill from one pixel to another. This color bleeding lowers the effective perceived resolution. As a result, rear-projection screens usually produce crisper pictures than front-projection screens.

It is important to consider the throw ratio of projectors when building projector arrays. To reduce the wasted space between the screen and the projectors in rear-projection systems, projectors of short throw distance are desirable. On the other hand, front-projection systems usually use projectors with larger throw distances. For many projectors, add-on short-throw and long-throw lenses can be purchased.

The zoom setting of the projectors is also an important factor. While radial and vignetting distortion exist in projectors, they are typically not noticeable in a single projector. Such distortion, however, becomes quickly noticeable when attempting to align multiple projectors. More importantly, these distortion effects vary with the zoom setting of the projector. Work by Yang et al. [103] found that for a variety of projectors, only a single zoom setting resulted in insignificant geometric barrel distortion. As a result, it is often desirable to set all the different projectors to that one "sweet" zoom setting. The other option is to devise camera-based methods to correct for the nonlinear distortions.

Projectors can be arranged in both abutting and overlapping configurations. Depending on the adversity of the vignetting effect, considerable brightness fall-off from the center to the fringes of the projector image could

Figure 2.11. Conception for Office of the Future. (Image courtesy of the University of North Carolina at Chapel Hill.)

be present. As a result, abutting configurations would show severe spatial brightness variation, especially for rear-projection systems. Our analysis shows that this vignetting fall-off can be as large as 40% of the peak brightness. So, the overlapping configuration is typically preferred, since it compensates for some of the radial fall-off along the projector fringes by overlapping projectors near the boundaries where their brightness properties are the worst. In addition, overlapping projectors alleviate the prob-

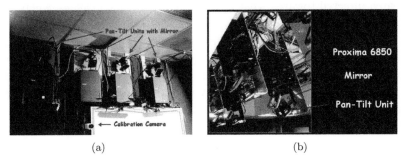

Figure 2.12. (a) Projection systems with mirrors on computer-controlled pan-tilt units for reconfigurable display walls at the University of North Carolina at Chapel Hill. (b) Zoomed-in picture of a single projector. ((b) © 1999 IEEE. Reprinted, with permission, from [80].)

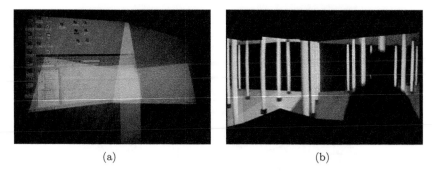

(a)   (b)

Figure 2.13. (a) Tiled displays not restricted to rectangular arrays. (b) Tiled display on a nonplanar surface. (Images courtesy of the University of North Carolina at Chapel Hill.)

lem of mechanical rigidity of projector arrangements. They can be moved around a little bit, and the software calibration techniques can detect this via a camera and take care of it. In particular, with slight movements in projector position and orientation there is no risk of creating a narrow region that is lighted by none of the projectors at all.

In fact, for some advanced applications, even the restriction of rectangular projection or planar displays can be prohibitive. For example, the Office of the Future application, as illustrated in Figure 2.11, proposes the use of the nonplanar walls of everyday office spaces as displays. The goal is to have "pixels everywhere," which can be used for something as complex as 3D visualization and teleconferencing applications or something as simple as our desktop. The US Department of Energy (DOE) is supporting recent research directed towards what is called *reconfigurable displays*, where the projectors can be moved around to create different types of displays based on the user's requirements. In such cases, mirrors mounted on computer-controlled pan-tilt units are used to move the projector's projection area around, as shown in Figure 2.12. Thus, it is possible to create displays that do not have projectors projecting only in a rectangular fashion, or only on planar displays, as shown in Figure 2.13.

2.6.2 Projector Considerations for Multi-Projector Arrays

There are various features and options in projectors that can be especially conducive to multi-projector display design. In this section, we discuss a few of the more important features that one might consider, especially when purchasing several projectors for use in a projector array.

Native resolution. As previously mentioned, the pixel array that makes up the panels in DLP and LCD projectors is physically fixed and therefore the device has a fixed number of output pixels, called its *native resolution*. While the projector can accept images of lower resolution than the native resolution, the input imagery must be resampled to fit the native resolution, which can result in undesirable sampling artifacts. Some projectors will even allow higher-resolution input than their native resolution. In this case, the input image is simply down-sampled to the native resolution. This can be a bit misleading to some users, who might think they are getting higher resolution than possible on their projector, but this is generally provided as a matter of convenience to avoid having to switch the device feeding the projector into a different resolution. To get the best image quality, it is recommended that the projector's native resolution match the resolution of the input image.

Video input (analog versus digital). Projectors are manufactured to accept either analog video, via a video graphics array (VGA) input, or digital video, specified by the digital visual interface (DVI) standard. Projectors accepting DVI will typically also accept the video in an analog format, but generally not vice versa. While analog video input has been the dominant format due to the long reign of CRT monitors, the trend is shifting towards digital. The advantages of digital are better expressed by the disadvantages of analog. For LCD and DLP projectors, the analog signal must be digitized into the individual pixels corresponding to the pixel array on the panels. While such analog-to-digital (A-to-D) conversion is performed automatically with typically good results, the conversion is not always perfect. As a result, projectors will often have controls that allow the user to manually tweak parameters used in the A-to-D conversion, in particular via controlling the phase of the the horizontal synchronization of the video signal to get the sampling correct. Such controls often come in two different scales, coarse and fine. The best way to see this effect is to display a pattern of alternating lines at the native resolution of the display. Banding effects due to poor A-to-D sampling are sometimes noticeable and must be fixed by adjusting the controls, shifting the phase.

Using the DVI format avoids the A-to-D conversion problems and should always result in a crisp picture. Note, however, that this requires the device feeding the projector to produce DVI output, so it is necessary to make sure that your PC or rendering hardware that drives the projectors is DVI-compliant. Most current high-end graphics cards, and increasingly even laptop computers, support the DVI format.

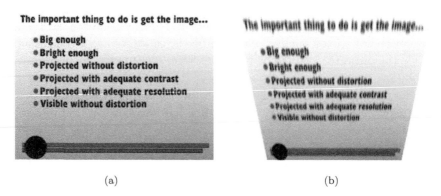

(a)   (b)

Figure 2.14. (a) On-axis projection with no keystoning. (b) Off-axis projection leading to keystoning.

Lens shift. The most common way to set up projectors is to place them on a table and use them for presentations. An on-axis projection will result in a rectangular image that will extend both above and below the tabletop level. Usually, projecting below the table level is undesirable. So, most projectors are set up for an off-axis projection, where the image is projected only above the tabletop level. When this is achieved without actually shifting the lens assembly, it introduces a keystone effect. This results in a trapezoidal image instead of a rectangular image, as shown in Figure 2.14. Most projectors provide a mechanism to correct for the keystone effect digitally. This requires image resampling and can thus result in sampling artifacts. However, many high-end projectors provide lens-shifting capabilities, where the entire lens system can be shifted, either mechanically or by hand. This allows the lens system to project a rectangular image even in the presence of a significant amount of off-axis projection. Thus, it avoids the need for digital keystone correction. While we will discuss techniques in the following chapters to allow for flexible positioning and avoid the need for lens shifting, there are still some advantages to having this ability. The most notable is focus. When a projector is significantly off-axis, it is possible that parts of the image are in focus while others are not, as shown in Figure 2.14. Lens shifting can help ameliorate this focus issue by shifting the optics.

Loop-through. Loop-through is the ability of a projector to have the input video signal loop through the device and come back out as output. This allows a video input to be fed from a PC to the projector and then back to a monitor. This feature can be very useful for diagnosis and maintenance of the PCs used to drive the projector array, in that the output of a PC attached to the display can be easily displayed back to a monitor without having to unhook it from the projector. If such a feature is only rarely needed, then a relatively cheap hardware video splitter can be used instead. Note that this feature is typically only supported for analog input and output. For projectors that support both VGA and DVI input, the loop-through output is only VGA and only loops the VGA input, i.e., DVI input will not loop through to VGA output.

Serial-port control. One may consider the availability of controlling the projectors via a serial-port connection, a feature provided by some manufacturers. When a very large array of projectors has been installed and housed, it can be quite tedious and time-consuming to make small adjustment to individual projectors manually. Having the ability to communicate with the projector over a serial connection can be a great benefit. Note, however, that this option should be carefully considered and explored with the projector vendor. While serial support is sometimes provided, it can be poorly documented, and it can be difficult to get detailed specs from the vendors. In addition, mechanical control (such as zoom and focus) can be disappointing, even when performed via serial control. For example, when controlling zoom and focus, serial commands can often only adjust the motors controlling zoom and focus up or down, in the same fashion that the user pushes the increase or decrease button. This means that setting the projector to an absolute zoom setting or focus setting is not possible even via the serial interface. One should also consider the additional cabling needed to achieve serial control; however, this can be done reasonably efficiently using hardware that supports virtual serial ports using IP.

Remote control. Many projectors come with remote controls that mimic the ability of the projector's on-board operational panel. This can be very useful when setting the controls of a projector that has been mounted out of reach. One thing to consider is that the frequencies used by these remote controls are typically the same for every projector of a particular make and model. This means that all the projectors in the array will respond to a single remote. While this can be very useful in powering on all the projectors, it can prove annoying when trying to adjust a single projector. Typically, when the remote is held close enough to an single projector, the signal will be directed to that projector alone, allowing for individual control.

2.6.3   Checklist of Important Considerations

Building a multi-projector display involves buying several projectors with
which the user needs to be happy for a considerable amount of time. Replacing projectors is not trivial. As we will see in later chapters, buying a
projector of a different make and model often introduces undesirable differences that can be difficult to handle. At present, the projector market
is advancing at such a fast pace that often the same model projector is not
available at a later date. It is therefore wise to check and double-check
the components before you finally decide to make a purchase of a large
number of projectors. When purchasing individual projectors, consideration is typically given to the price, resolution, and size. This is reflected
in reviews of projectors found on the Web and in trade magazines. When
buying several projectors, other considerations like noise and heat, which
are often overlooked when buying a single projector, are also important.

	Projector A	Projector B	Projector C
From Spec Sheet			
Price			
Type (DLP, LCD, CRT)			
Lumens			
Contrast ratio			
Native resolutions			
Lens FOV (zoom range)			
Analog or digital input			
Weight			
Projector controls provided			
Brightness/contrast			
Color balance			
etc.			
Lens shift available			
Warranty (num. of years)			
Noise level			
Heat output			
Lamp life			
Replacement lamp cost			
Video loop-through			
Field Test			
Overall response			
Start-up time			
Design			
If two projectors are available			
Color consistency b.t. projectors			
Red			
Green			
Blue			

Figure 2.15. Specifications to consider when comparing projectors.

Figure 2.15 is an example of specifications that one may consider when comparing projectors. These are divided into two categories: *product specifications*, which encompasses those specifications that can be easily obtained from product specification sheets, and *field-test* properties, which require the projector to be tested before purchasing. In the case where several projectors will be purchased, vendors will often let the customer try out the projector beforehand. If this option is possible, it is highly recommended that you request to try out two projectors of the same model. This way, color consistency between projectors can be tested.

2.6.4   Driving Architecture

Outside the projector set-up, generating the imagery for the projector array is the other key component of designing a large-scale display. The requirements of this particular component are tied tightly to the specific task at hand and the type of data being displayed. While such rendering was once only possible through very high-end specialized rendering computers such as Silicon Graphics machines, the trend now is toward the use of clusters of PCs, in particular, PCs with high-end (yet affordable) graphics cards.

In this book, we will assume a PC-cluster driving architecture, with one dedicated PC per projector. While there can be many variations on the rendering architecture, the most common is where a *centralized* server is used to manipulate the content of the displays, either in terms of images or a location inside a 3D scene. This centralized server communicates with the individual PCs driving each projector, controlling the content to be displayed. Such an architecture is illustrated in Figure 1.4. As previously mentioned, the amount of computing power and associated resources, e.g., memory and backing store, as well as the network communication speed, are specific to the task of the display.

Accessing the PC cluster for set-up, maintenance, and software set-up will be necessary. While remote-desktop applications can be used in many cases, more often than not, the PC must be accessed directly. A wise investment is a keyboard and monitor switch such that only one keyboard and monitor is necessary for the cluster. The need to use a monitor (or, instead, simply work from the associated projected imagery) is dependent on the set-up and amount of overlap from the other projectors. If a monitor is needed, then consideration should be given to a video repeater or loop-through options on the projectors. Another consideration is the cabling of all the projectors and PCs. We highly recommend investing in a high-quality label maker for labeling each cable and its associated PC.

While PC clusters make projector arrays more affordable, we note that a PC cluster with a centralized server does have some limitations. It is relatively difficult to scale (increase the number of pixels) and reconfigure (change the aspect ratio) the display with a centralized architecture. However, this is the current state-of-the-art architecture, and all of the calibration techniques for the display, both in terms of geometric and photometric issues, assume such an architecture. As a result, these set-ups typically use a camera that is attached to the centralized server controlling the individual PCs and corresponding projectors. However, the limited resolution of a single camera can become problematic when calibrating a very-high-resolution display. To address this issue, multiple cameras are often required, but managing multiple cameras with a single centralized server poses its own scalability problems and is difficult. To address all of these issues, recent research is exploring the use of a *distributed* architecture via a network of projector-camera systems for scalable and reconfigurable displays. The concluding chapter of this book, Chapter 6, is dedicated to discussing this new architecture.

2.7 Summary

Our experience with multi-projector displays reveals that it is immensely useful to have a clear understanding of all of the elements of projection-based displays, the causes of artifacts, and the parameters used for characterizations and how they are calculated. These help in targeting devices that are most conducive to generating large displays. In this chapter, we have presented a brief but comprehensive treatment of all of these aspects. The goal was to familiarize the reader with several trade-offs and options presented so that he is educated to make an intelligent choice.

However, striking a balance between all of these features and configurations is often not easy. At some point, a decision will have to made as to the projector make and model or projector configurations that are best suited for a particular display set-up. Projector technology is advancing at a rapid pace, which means quality and features are continually changing. There are many websites now with comprehensive lists of projectors, their basic specifications, and often reviews. While Web-based reviews are typically not complete in all the issues one may want to consider in a projector array, they are a nice starting point and can help navigate through the mind-boggling number of models and options.

3

Geometric Alignment

A S DISCUSSED IN CHAPTER 2, a key challenge in building a tiled projection-based display is to ensure that image contributions from multiple projectors align on the display surface to create an overall image
that appears correct and seamless. This task is challenging for projector-
based displays because the image-generation hardware is not integrated
with the display surface. This alignment is traditionally achieved by physical construction, where projectors are mounted on sophisticated control-
lable mounts that are part of the display infrastructure. Even with these
projector mounts, subsequent manual adjustment is typically needed to
refine the projectors' position and orientation and to ensure an accurate
alignment of the projected imagery.

In this chapter, we discuss how geometric alignment can be achieved
by warping the projected imagery instead of physically adjusting the projectors' position and orientation. This image-based warping technique is
already used in a limited fashion in current projectors. For example, almost
all projectors have some built-in keystone correction that can warp the input image to compensate for distortions caused by off-axis projection. This
warping to correct for keystoning is typically applied along the horizontal
scanline only, assuming the most common case of presentation where the
projector is placed on a table and keystoning only affects the horizontal
scanlines. The techniques discussed in this chapter involve the same basic
concept but use more flexible warping methods that can assure alignment
with neighboring projectors and even compensate for projection onto nonplanar surfaces. Key to the success of these flexible warping approaches is
camera-based feedback that uses one or more cameras to observe projected

features from each projector in the multi-projector array and computes the
necessary warps to ensure both geometric alignment and correctness.

This imaged-based approach to projector alignment has been referred to
by many names, including *geometric correction, image pre-warping, post-render warping*, and *software alignment*. In this book, we will refer to it
as *geometric correction*, and we will refer to the camera-based techniques
that compute the necessary parameters as *geometric registration*.

We also refer readers to an excellent companion book by Bimber and
Raskar that discusses related geometric registration and correction for
projector-based applications outside of tiled displays [8].

3.1 Nature of the Display Surfaces

Several camera-based approaches have been introduced for geometric registration of multi-projector displays (e.g., see [2, 12, 19, 20, 23, 36, 78, 79, 80, 83, 95, 96, 97, 102, 103, 13]). These different approaches can be
classified primarily by the type of display surface they address: *parametric*
and *nonparametric*. For parametric surfaces, the display surface can be
assigned a 2D parameterization that has a direct relationship with the parameterization of the image to be displayed; most notable in this category
are planar display surfaces. Nonparametric surfaces are those for which a
suitable parameterization does not exist or is not established. Approaches
that handle these two different types of surfaces are similar in their goal of
producing a seamless image from multiple overlapping projectors but are
fundamentally different in the way they achieve it.

3.1.1 Parametric Display Surfaces

Geometric registration techniques for parametric display surfaces target
projection-based displays that generate imagery on display surfaces that
can be easily parameterized. It is assumed that the display-surface parameterization has a direct relationship with the parameterization of the
image to be displayed. While parametric approaches are most often used
with planar display surfaces, they can also be applied to a parameterizable
surface such as cylindrical or spherical displays.

Consider a planar display surface D that is parameterized by normalized
parameters (s', t') along the horizontal and vertical directions, as shown in
Figure 3.1. Also assume that the image I to be displayed is parameterized
with different parameters (s, t). For simplicity, we assume that the image

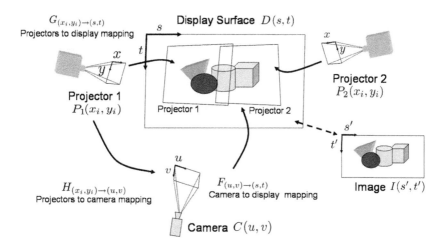

Figure 3.1. An example of a parametric display surface and associated coordinate frames and mappings. Examples of approaches targeting this display type can be found in [97, 19, 83, 20, 23, 36].

and the display have the same parameterization, i.e., $(s', t') = (s, t)$. While an identical correspondence is typically not the case, the mapping between a planar display surface and the image is generally trivial, typically involving only a scaling and translation.

To produce the image I on the display surface D, it is necessary to determine how the individual projectors' images map to the image I. Let the display be composed of N projectors. Each projector P_i, $1 \leq i < N$, has its own coordinates (x_i, y_i). The goal, then, is to determine how each projector's coordinates (x_i, y_i) map to the image coordinates (s, t). Since the display D coincides with the image I in a parametric display surface, this mapping automatically provides the mapping between the projector coordinates and the display coordinates. This mapping between the individual projectors and the image is expressed as $G_{(x_i,y_i)\to(s,t)}$. To determine G, a camera (or possibly multiple cameras) is introduced into the environment. The camera C has its own coordinates (u, v) and can observe the display surface, establishing a mapping from its coordinates to the display coordinates and, hence, to the image coordinates (s, t). This mapping is denoted by $F_{(u,v)\to(s,t)}$. Projected features from each projector can be observed by the camera, establishing a mapping from the projector coordinates to the camera coordinates, denoted by $H_{(x_i,y_i)\to(u,v)}$. The final mapping G (from projector to image) is achieved by a concatenation of the mappings H (from the projector coordinates to the camera coordinates) and F (from the camera coordinates to the image coordinates). Thus,

$$G_{(x_i,y_i)\to(s,t)} = F_{(u,v)\to(s,t)} \cdot H_{(x_i,y_i)\to(u,v)}. \qquad (3.1)$$

The key idea of the approaches that target parameterized display surfaces is that the display surface's parameterization is exploited to provide a common coordinate frame in which the projector, camera, and image coordinates can be canonically expressed. The camera's role is essentially to provide a mechanism to establish the mapping between these coordinate frames. In these approaches, "geometric correctness" means geometric alignment of projected imagery on D. The result of this approach is a virtual image created from contributions from multiple projectors that appears as if the image I is wallpapered correctly (i.e., aligned) onto D to create an image with no geometric discontinuities.

3.1.2   Nonparametric Display Surfaces

Approaches dealing with nonparametric displays are applicable to projector arrays with display surfaces that cannot be easily parameterized or for which a parameterization is not desired. Figure 3.2 shows an example of a

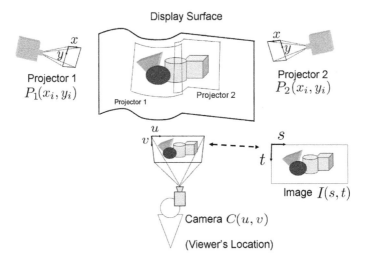

Figure 3.2. An example of a nonparametric display surface and associated coordinate frames and mappings. Examples of approaches targeting this display type can be found in [79, 80, 12, 103, 102, 78].

display surface D where there is no direct relationship between the display D and the image to be displayed I. For displays using this type of surface, it is not possible to exploit the display's parameterization as a common coordinate frame to describe the various devices.

Instead, the goal of these approaches is to find a mapping for image I that maps directly into the *viewer's eye*. A camera is placed at the desired viewing location to mimic the viewer. The camera is used to capture a reasonable approximation of how the light rays from the projectors form an image on the viewer's retina. This is achieved by assuming a direct correspondence between the camera coordinates (u, v) and the image coordinates (s, t). This correspondence is often just a scaling and translation. Thus, in this approach, there is no need for an explicit mapping from the camera to the display. Instead, we only need to compute the mapping, $H_{(x_i, y_i) \to (u,v)}$ that maps each projector's $P_i(x_i, y_i)$ to the camera image $C(u, v)$. As a result, Equation (3.1) becomes

$$G_{(x_i, y_i) \to (s,t)} = H_{(x_i, y_i) \to (u,v)}.$$

All that is required for nonparametric displays is to define the relationship G from the projector coordinates to the image coordinates such that the projected images create the image I when captured by the camera C. The mapping G is estimated by projecting features with known locations (x_i, y_i) in the projector, which are then observed by the camera. The image displayed on the nonplanar surface looks correct to a viewer placed at the location of the camera.

This approach is quite powerful because it does not require any knowledge of the display surface. The mapping $H_{(x_i, y_i) \to (u,v)}$ implicitly encodes the display surface's shape, the projector orientation, and even the projector radial distortion. Raskar et al. [79] explains this correction technique as analogous to having a slide projector positioned at the viewer's location projecting the desired image I onto the display surface. The display surface illuminated by this image is then rendered from each projector's point of view. These images rendered from the projectors' points of view are the ones that need to be displayed by the projectors to produce the correct desired image I to the viewer.

While having a similar goal, the approaches for handling parametric and nonparametric surfaces are fundamentally different. This is best seen by using the nonparametric technique on a planar display surface as shown in Figure 3.3. For the parametric display surface, the goal is for the projected imagery to produce an image aligned on the display surface in a wallpaper-

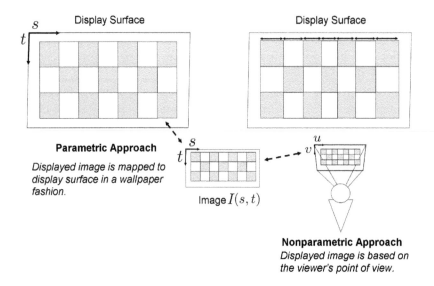

Figure 3.3. Differences between parametric and nonparametric approaches.

like fashion. For the nonparametric approach, the goal is to have an image that looks like the image I from the location of the camera, which is the intended viewing location. Figure 3.3 shows an example where a checkerboard pattern is being displayed using the two different approaches. The nonparametric approach will produce a checkerboard pattern on the display surface where the corners of the checker-patterns may appear unequally spaced to compensate for perspective distortion (this effect has been *greatly* exaggerated in the Figure 3.3). However, when viewing this unequally spaced checkerboard pattern from the intended viewing location, it will appear as an equally spaced checkerboard pattern.

Note that both of these methods will produce images that are geometrically aligned at the projector boundaries and are perceptually seamless. The question may arise as to which is better or more correct. The answer depends entirely on the application at hand, but in general, both produce a desirable display. In the parametric case, although the image does not look perspectively correct from any particular viewing location, this wallpaper effect is very familiar to viewers. As long as the image I appears wallpapered correctly onto the display surface, viewers generally have no problem accepting it. This is especially true for planar display surfaces.

On the other hand, producing an image on the display that compensates for our viewing location is also acceptable and is a technique often employed in art and architecture. Michelangelo's *David* statue was carved with a disproportionately large upper body and head to compensate for perspective distortion caused by the viewer's having to look up from the ground. Adornments drawn on the Taj Mahal appear uniform in size even as they "climb up" the high walls; this is because the adornments increase in size with altitude on the wall to compensate for perspective distortion. The limitation of this approach, however, is that the correctness is tied to the viewing location. As the viewer moves away from the intended viewing location, often called the "sweet spot," the image will appear distorted. However, as with the wallpaper effect, viewers are generally familiar with looking at an image from a slightly different perspective without noticing the artifacts. This, of course, is a function of the display surface and viewing location, but it is often not an issue if the display surface is not unnaturally distorted.

Deciding on which approach to use in a large-scale display design is often not a matter of the difference in the final imagery but of the manner in which the mappings F and H can be established and the final warping G applied to the image. This is the topic of the rest of this chapter.

3.2   Geometric Registration Techniques

All geometric registration techniques try to estimate the function G as accurately as possible. However, they differ in the kinds of functions (linear, piecewise-linear, or nonlinear) that are used to represent G and its components F and H. The choice of these functions is tied closely to the type of display surface, the amount of geometric imperfections present in the projectors, and the desired accuracy of the geometric registration.

The most common case is to use a single camera in a centralized fashion where a single computer (or cluster of computers operating as a single computer) controls all of the projectors and the camera in order to estimate the function G. A single high-resolution camera is often sufficient for small or medium displays ranging anywhere from four to 20 projectors. However, as the scale gets larger, multiple-camera techniques can be used on limited types of display surfaces. In this chapter, we discuss both single- and multiple-camera techniques. In Chapter 6, we discuss some recent work on scalability displays that allows a plug-and-play usage of projectors.

We first discuss the simplest case of planar parametric displays. Next, nonplanar parametric displays are discussed, followed by the most general case of arbitrary nonplanar nonparametric displays.

tem D via a linear homography. This projector-to-display homography is denoted by $\mathbf{H}_{P_i \rightarrow D}$ and maps the image of P_i to the reference display coordinate D. The goal is to estimate for each projector P_i the $\mathbf{H}_{P_i \rightarrow D}$ using a single or multiple cameras.

Single-camera approach. We first consider the case where only one camera is used as shown in Figure 3.5. A homography between the camera C and the display D, denoted by $\mathbf{H}_{C \rightarrow D}$, is first computed. This is typically performed by placing physical markers on the actual display surface forming the reference frame for D. Point correspondences between the physical markers and the observing camera are established either manually or using feature-detection algorithms. From these correspondences, the homography $\mathbf{H}_{C \rightarrow D}$ is computed.

After $\mathbf{H}_{C \rightarrow D}$ has been derived, projected imagery from each projector P_i is observed by the camera C, and a projector-to-camera homography $\mathbf{H}_{P_i \rightarrow C}$ for each projector P_i is calculated. The projector-to-display homography $\mathbf{H}_{P_i \rightarrow D}$ is then computed from $\mathbf{H}_{P_i \rightarrow C}$ and $\mathbf{H}_{C \rightarrow D}$ as

$$\mathbf{H}_{P_i \rightarrow D} = \mathbf{H}_{C \rightarrow D} \times \mathbf{H}_{P_i \rightarrow C}, \tag{3.2}$$

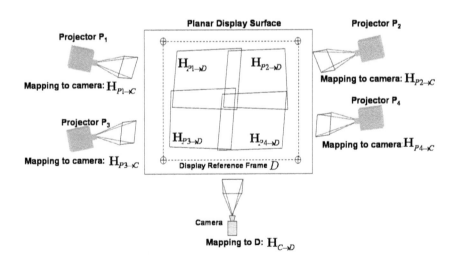

Figure 3.5. A single camera is used to establish a homography between the projectors and a reference frame on the display surface. Examples using this simple set-up can be found in [20, 102, 81]. (© 2005 IEEE. Reprinted, with permission, from [13].)

Figure 3.6. Results from the UNC PixelFlex system [102]. This 4×2 projector array has used a single camera to compute the necessary homographies to align the imagery. (© 2001 IEEE. Reprinted, with permission, from [103].)

where the operator $\times$ represents a matrix multiplication. As previously described, the inverse homograpies $\mathbf{H}_{D \to P_i}$ define the warps to be applied to the image I to produce the appropriate image to be projected by each projector.

For reasonably sized display walls that can be easily viewed by a single camera, this technique can be effective and fast. Raskar et al. [81] demonstrates a system that uses this technique and computes the necessary homographies for a 2×2 projector array in a few seconds. Figure 3.6 shows a 4×2 array of eight projectors that has been registered using this method.

Multiple-camera approach for planar displays. The ability of homographies to be concatenated makes this method suitable to a multiple-camera approach. This approach is useful for very large projector displays where there is a significant mismatch in the resolution of the camera and display, resulting in unacceptable errors in the geometric alignment. In an effort to accommodate larger display resolutions and to utilize cheaper cameras with smaller fields of view and lower resolution, H. Chen et al. [20] propose a scalable homography-based technique that uses multiple cameras.

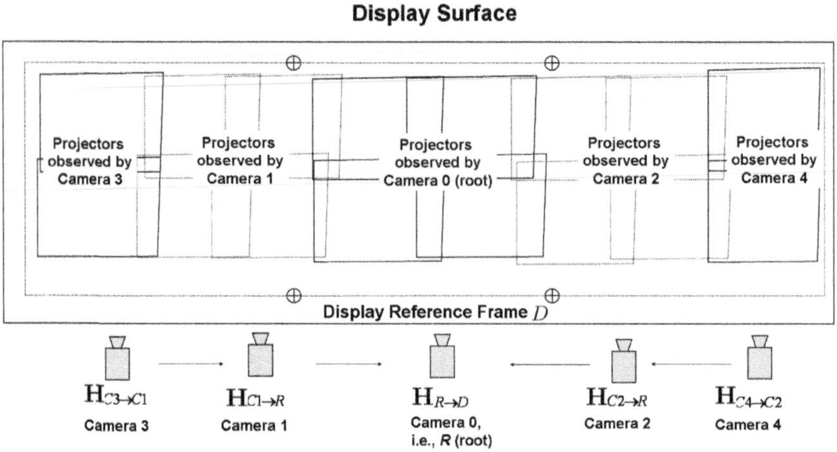

Display Surface

Figure 3.7. Multiple cameras are used to establish a series of homographies between the projectors and a reference frame on the display surface [20].

In this case, it is assumed that adjacent cameras have an overlapping field of view and see some common regions on the display. Adjacent cameras C_j and C_k are related to one another by the homographies $\mathbf{H}_{C_j \to C_k}$ since they observe the same planar display. Point correspondences are established between adjacent cameras by observing projected points in their overlapping field of view, and the $\mathbf{H}_{C_j \to C_k}$ are computed from these correspondences. Note that $\mathbf{H}_{C_j \to C_k}$ can be inverted to represent the inverse relationship $\mathbf{H}_{C_k \to C_j}$. Mathematically,

$$\mathbf{H}_{C_k \to C_j} = \mathbf{H}_{C_j \to C_k}^{-1}.$$

Next, a root camera R is chosen, and the homography relating this root camera to the display, $\mathbf{H}_{R \to D}$, is computed. Each of the other cameras C_j are registered to the root camera and hence to the display D by a homography denoted by $\mathbf{H}_{C_j \to D}$. This homography is constructed by concatenating adjacent camera-to-camera homographies until the root camera is reached, as follows (see Figure 3.7):

$$\mathbf{H}_{C_j \to D} = \mathbf{H}_{R \to D} \times \mathbf{H}_{C_k \to R} \times \cdots \mathbf{H}_{C_j \to C_k}. \tag{3.3}$$

The homography $\mathbf{H}_{C_j \to D}$ maps coordinates in camera C_j to the display D. To determine the path of this camera-to-display concatenation, a minimum spanning *homography tree* is built that minimizes registration errors across adjacent cameras [20].

To estimate the final mapping between each projector and the display through the cameras, each projector is related to the camera(s) observing it. A single camera in the set-up typically observes only two to four projectors. The projector P_i can be related to its corresponding camera(s) C_j via a homography, denoted by $\mathbf{H}_{P_i \to C_j}$. Using the homography tree computed between the cameras, the projector-to-display homography $\mathbf{H}_{P_i \to D}$ for a projector P_i can then be computed as

$$\mathbf{H}_{P_i \to D} = \mathbf{H}_{C_j \to D} \times \mathbf{H}_{P_i \to C_j},$$

where $\mathbf{H}_{C_j \to D}$ is constructed using Equation (3.3).

This approach is similar to methods used for building large image mosaics with images coming from multiple cameras and is very much akin to traditional image-mosaicing techniques (e.g., Shum and Szeliski [88]). The constructed image mosaic from the multiple cameras is large enough to observe all of the projectors and establish their relationship to the display surface. Moreover, the projector-camera environment aids the image-mosaic construction because correspondences between cameras can be accurately obtained by observing the projected features.

However, when using multiple cameras, accumulation of errors along a path in the homography tree sometimes results in an error of a few pixels across projector boundaries. Usually, each homography in Equation (3.3) involves an estimation error. Since multiple homographies are concatenated to create the final projector-to-display homography, the errors can accumulate and result in a misalignment of a few pixels. The errors can be reduced by using global error diffusion methodologies resulting in local alignment of subpixel accuracy [20].

3.2.2 Nonlinear Method for Parametric Planar Display

The underlying key assumption of the homography-based approaches is that all devices behave linearly. This is often not the case. Most commodity cameras exhibit lens distortion. Even expensive high-resolution cameras, as are mostly used in geometric registration of multi-projector displays, show some small lens distortion. These can be easily calibrated using widely prevalent calibration techniques, some even available online [9]. Projectors also exhibit radial distortion (details on radial distortion correction are presented in Appendix C). While for relatively expensive projectors lens distortion can be virtually undetectable, this is not the case for inexpensive commodity projectors. Projector lens distortion can be more significant for projectors with a short throw distance, due to their short focal length. This distortion problem is especially prevalent in rear-projection displays, which are forced to use such short-throw projectors. Thus, the pure homography-based approaches may introduce errors as these nonlinearities become more significant or compound. A study by Yang et al. [102] shows that the zoom settings of the projectors can be pre-manipulated to a "sweet spot" where the projectors show the minimal radial distortion; it is important to set the projectors to these settings before applying any homography-based approach.

Hereld et al. [38] proposes a method for such scenarios where the projectors are related to the camera using a nonlinear cubic polynomial. Thus, $H_{(x_i,y_i) \to (u,v)}$ in Equation (3.1) is estimated using cubic polynomials instead of a linear homography. Without loss of generality, we denote (x, y) as the projector pixels and express its relationship to the camera pixels (u, v) as follows:

$$
\begin{aligned}
u &= a_0 + a_1x + a_2y + a_3xy + a_4x^2 + a_5y^2 \\
&\quad + a_6x^2y + a_7xy^2 + a_8x^3 + a_9y^3, \\
v &= b_0 + b_1x + b_2y + b_3xy + b_4x^2 + b_5y^2 \\
&\quad + b_6x^2y + b_7xy^2 + b_8x^3 + b_9y^3.
\end{aligned}
\tag{3.4}
$$

Unlike homographies, nonlinear functions are not easily invertible. Hence, tree-like techniques akin to homography trees that can scale to multiple cameras cannot be realized using such nonlinear methods. As a result, this method is applicable only to relatively smaller displays where all of the projectors are easily visible from a single camera.

Just like the homography-based method using a single camera, each projector is first related to the camera by a nonlinear function (instead of a linear function). The camera, however, is related to the display by a linear function since the presence of a planar screen lends itself to an accurate linear representation. These two mappings are used to extract the appropriate part of the image I for each projector P_i.

This method can help to avoid manual presetting of projectors to the "sweet" zoom level, especially for commodity projectors, and can also yield a compact representation of the warping parameters as a few coefficients of the cubic polynomial. However, data fitting is required to estimate a large number of parameters, and it requires a dense set of correspondences between the projector and the camera to calculate the 20 coefficients

(a)

(b)

Figure 3.8. A 3 × 3 projector display wall registered using nonlinear warping. (a) Before registration. (b) After geometric registration using a nonlinear method [38]. Photometric registration is not applied.

$(a_i, b_i, 0 \leq i \leq 9)$ of the cubic polynomial [38]. This method usually yields subpixel accuracy in the presence of small lens distortion in the projectors. Figure 3.8 shows an example of a 3 × 3 projector display wall that has been registered using the nonlinear parameterization of the warping function.

3.2.3   Piecewise-Linear Method for Parametric Planar Display

Both methods discussed so far for planar displays take the approach of representing the maps H and F using parametric functions. Estimating parametric functions requires data-fitting computation, and errors are related to the quality of this estimation. Although these result in overall reasonably aligned and correct imagery, local errors can result in small local misalignments. For most imagery, this may not be noticed, but for some cases, such as when displaying text, this may be visible. A complementary technique is to avoid parametric functions altogether and use a piecewise-linear approach.

For imperfect devices, such as projectors with lens distortion, the mapping H is nonlinear. The approach described here uses a piecewise-linear approximation to estimate this function instead of a compact parametric representation. To acquire this piecewise approximation, a relatively dense correspondence between the projector coordinates and the camera coordinates is first established. This is achieved by projecting equally spaced features from the projector and detecting their positions in the observed camera image. Again, without loss of generality, let us assume the projector's coordinate system to be defined by (x, y) and the camera's by (u, v). Let the number of such correspondences be m and denoted by $(u_k, v_k) \rightarrow (x_k, y_k)$, $0 \leq k < m$. The (u_k, v_k) are tessellated to create a triangulation in the camera coordinates, denoted by T_C. The (x_k, y_k) in projector P corresponding to (u_k, v_k) are connected in exactly the same manner as T_C to generate the corresponding triangulation in the projector's coordinate space. This corresponding triangulation for each projector P_i is denoted by T_i. The T_i and T_C thus encode the mapping H between the projector and the camera.

The mapping F between the camera and the display can still be achieved by a compact homography due to the presence of the planar screen. Warping an image based on this piecewise mapping is described in Section 3.3.4. Figure 3.9 shows an example of the projected features and corresponding triangulated mesh generated in the camera space.

This piecewise-linear method allows nonlinear warping to be realized on parametric display surfaces. This method works and can easily achieve subpixel accuracy, even in the presence of some lens distortion in the projector. It relies on a piecewise-linear approximation of the function relating the projectors and camera and hence avoids any errors resulting from data fitting. However, it is not easily scalable to multiple cameras. Further-

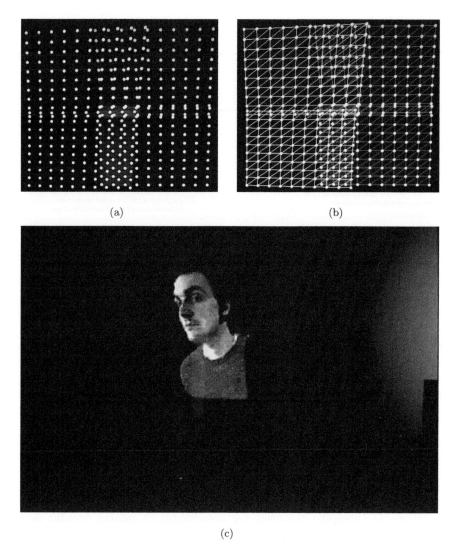

Figure 3.9. Piecewise approach used on a planar display surface. (a) Projected features from four different projectors. (b) The corresponding tessellated mesh. (c) The display in use. (© 2002 IEEE. Reprinted, with permission, from [12].)

more, the triangulated mesh T_i does not yield a compact representation, especially compared to a 3×3 homography. However, this is a relatively low cost to pay given the large amount of memory easily available in most specialized hardware today.

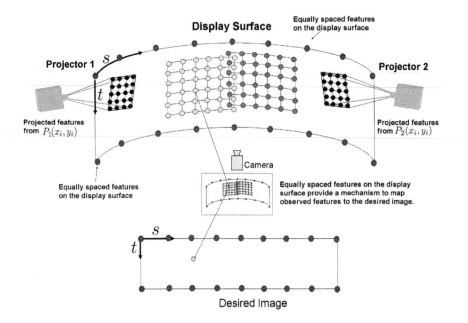

Figure 3.10. Work in [36] placed equally spaced fiducials along the top and bottom of the display surface. The display surface can be parameterized by (s', t'). Since (s', t') is orthogonal, it can be easily related to the image coordinates (s, t). The camera observes these equally spaced features and builds the function $F_{(u,v)\to(s,t)}$.

3.2.4 Piecewise-Linear Method for Parametric Nonplanar Display

Researchers at HP Research in Palo Alto recently proposed a method that can easily parameterize a cylindrical display [36]. In this approach, the functions H (mapping the projector coordinates to camera coordinates) and F (mapping the camera coordinates to display coordinates) in Equation (3.1) are both represented by a piecewise-linear transformation. Figure 3.10 shows a diagram of their technique.

As in the case of planar displays, a camera is used to observe the projected features on the display from each projector to generate the triangulations T_C and T_i, respectively. In addition, the curved display parameterization is achieved by equally spaced markers placed on the top and bottom of the display. The display is considered to be parameterized by s' and t' running along the horizontal curve and vertical axis of the curved section, respectively. These equally spaced physical markers that make up the top

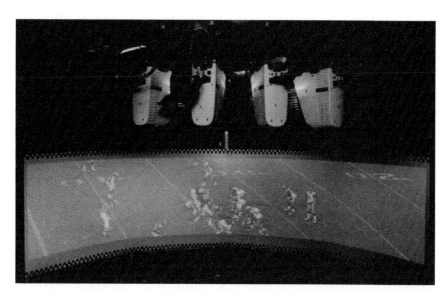

Figure 3.11. Example of the set-up described in [36]. The checkerboard pattern at the top and bottom of the display surface provides the necessary information to map the camera pixels to the display surface. (Image courtesy of Hewlett Packard Labs Palo Alto.)

Figure 3.12. Example of the set-up described in [36] without (top) and with (bottom) geometric correction enabled. (Images courtesy of Hewlett Packard Labs Palo Alto.)

and bottom curve of the display surface can be observed by the camera and detected in the camera coordinates yielding a set of correspondences between the camera coordinates and the display coordinates. Let there be n correspondences denoted by $(u_j, v_j) \rightarrow (s'_j, t'_j)$, $0 \le j < n$. The (u_j, v_j) are first tessellated to create a triangulation in the camera space T_C. The corresponding (s'_j, t'_j) in the display space are then connected in exactly the same way to create a corresponding triangulation in the display space T^D. Using these correspondences, a display triangulation can be constructed in the camera space, denoted by T_C^D. Now, any projector pixel (x_i, y_i) can be mapped to the display coordinates via a series of triangulations $T_i(x_i, y_i) \rightarrow T_C(u, v) \rightarrow T_C^D(u, v) \rightarrow T^D(s', t')$.

The use of the equally spaced markers applied on the display surface allows a direct link between the image $I(s, t)$ and locations on the display surface D and their corresponding pixel contributions from the multiple projectors. Like the planar-display approach, this mapping produces a geometrically aligned image that is wallpapered on the display but not necessary geometrically correct for a particular viewing position.

Figure 3.11 shows an example of the actual display set-up. The checkerboard pattern provides the equally spaced features along the top and bottom of the display surface. This provides the cues for the parametrization of this surface in the camera coordinates. Figure 3.12 shows the result with and without the geometric correction turned on. The geometrically corrected image looks correct and accurately aligned.

3.2.5   Nonparametric Display Surfaces

As discussed in Section 3.1.1, to register the projected geometry for a non-parameterized display surface, we need to find how the light from the multiple projectors maps to the viewer's eye. Essentially, the mapping F from the camera to the image in Equation (3.1) is simply ignored (or can be considered to be the identity), and an accurate estimation of H from the camera position is required. To achieve this, a camera is placed at the location where the viewer is intended to observe the displayed imagery. A set of equally spaced features are projected from each projector P_i and registered in the camera image plane. The projected features (x_i, y_i) are used to form a tessellated grid in the projector coordinates T_i and in the camera coordinates T_C, as shown in Figure 3.13. This establishes a nonlinear mapping from the projector's features (x_i, y_i) to their positions in the camera's image plane (u, v) via a piecewise-linear approximation denoted by $T_C(u, v) \rightarrow T_i(x_i, y_i)$. Note that in this case of nonparameterizable

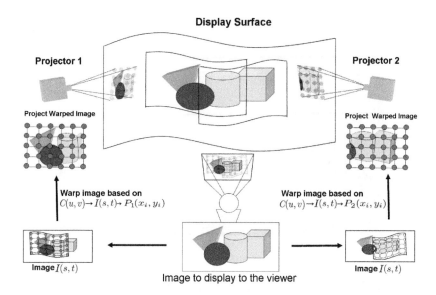

Figure 3.13. Geometric correction is achieved by piecewise warping of the desired image $I(s,t)$ to its corresponding projector pixel locations $P(x,y)$ based on the camera observations. Examples of approaches using this type of geometric correction can be found in [79, 80, 12, 103, 102, 78].

displays, the triangulation automatically encodes the geometry of the display surface without explicitly parameterizing it. Hence, the display can be completely ignored in the mappings.

To correct the displayed imagery, we can map the camera image coordinates (u,v) to the image coordinates (s,t) directly using the piecewise-linear mapping $T_C(u,v) \rightarrow T_i(x_i,y_i)$. This achieves the appropriate warping of image I that, when projected by each projector onto the arbitrary nonplanar display, results in an image that is geometrically correct from the camera position. Hence, the camera is positioned close to where the viewer will be positioned when viewing the display, as shown in Figure 3.13.

Figure 3.14 shows an example of this approach applied to a three-projector array projecting onto a curved wall in an atrium. A camera has been placed directly in front of the display. The imagery therefore is correct for a viewer standing in front of the display. Figure 3.14(b) shows the slight distortions in the image when viewing from a different spot other than the "sweet spot." The imagery still looks correct. Typically, the viewer can deviate several meters from the viewing location before significant distortion is noticed.

(a)

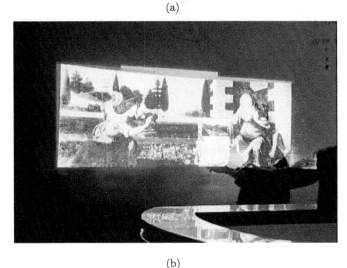

(b)

Figure 3.14. (a) Three-projector array on a curved surface. (b) Geometrically correct imagery on the display surface with projector overlap. The geometry is actually correct for the viewer's point of view; however, it is difficult to tell this from even this viewing location. (© 2002 IEEE. Reprinted, with permission, from [12].)

Note that this approach is fundamentally different from the parametric approach for a cylindrical surface. In the parametric case, the image will appear wallpapered onto the display surface, whereas in this nonparametric case, it will appear as if a perspective projection of the image is viewed from the camera's position. While fundamentally different, most viewers do not notice any difference between these two approaches unless explicitly asked to observe a difference.

3.3   Implementation Details

In this section, we describe implementation details of the different geometric-registration techniques. Some of the steps used, such as feature detection of projected patterns, are common to all geometric-registration techniques. However, others, such as determining the coefficients of the parametric functions or generating a valid triangulation, are applicable to a subset of the techniques.

3.3.1   Projected Patterns

Establishing correspondences between the projectors, display surface, and camera is the heart of the camera-based registration techniques. This is done by projecting patterns of known features that are subsequently observed by a camera or multiple cameras. Several different types of features have been demonstrated, all with similar results. Examples include a set of equally space circles or Gaussian blobs, where the centers of the blobs or circles are taken to be the known (x_i, y_i); checkerboard patterns, where the corners are used as the known (x_i, y_i); or the intersection of projected vertical or horizontal lines to form the known (x_i, y_i). Such patterns in the projector coordinate space are shown in Figure 3.15. These features are observed in the camera and identified, generating a relatively sparse sampling of the function $H_{(x_i, y_i) \leftrightarrow (u,v)}$.

3.3.2   Detecting Patterns

To detect features reliably, the projection of imagery should be synchronized with camera capture. This is followed by simple image processing to locate salient features such as lines and circles. Details of specific image-processing routines to find and localize observed features is outside the scope of this book, and we direct interested readers to any good book on image processing (for example, [35]).

The bigger challenge is to determine correspondence between the projected feature and detected feature in camera coordinates that immediately follows the feature-detection step. If the orientation of the projector's image on the camera is known, establishing the correspondence between the

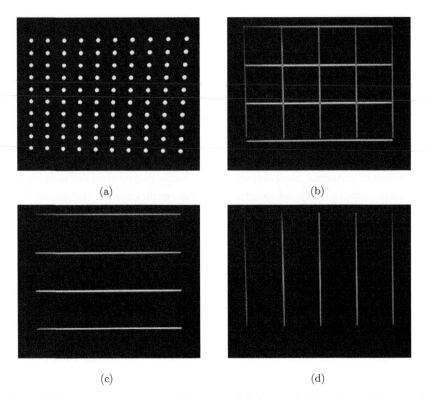

(a)   (b)

(c)   (d)

Figure 3.15. Example of projected features. (a) A set of 10×10 equally spaced circles as observed by a camera. (b) In this example, the features are the intersections of (c) horizontal and (d) vertical lines. The intersection points are the known $P(x_i, y_i)$.

features and camera is generally straightforward. The challenge is when the projector orientation is unknown (e.g., it could be upside-down or rotated) or if projecting on an arbitrary surface where the positions of the features are difficult to predict. One way to aid the identification of correspondences for these more challenging cases is to use a binary encoding scheme that uses temporal multiplexing of projected imagery.

When using binary encoding, every feature is assigned a binary-encoded ID, and the total number of frames projected corresponds to the total number of bits required to encode the IDs of the total number of features. Each projected frame corresponds to one bit plane of the feature IDs, and each feature is only turned on in the frame corresponding to the bit plane where its ID has a value of 1. If the total number of features is M, this scheme needs $\log(M)$ frames.

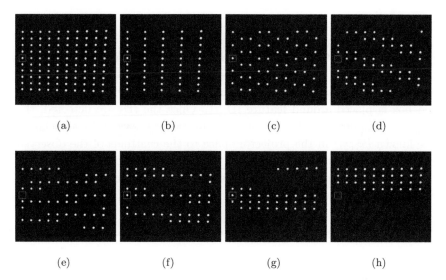

(a)	(b)	(c)	(d)
(e)	(f)	(g)	(h)

Figure 3.16. Projector-to-camera mapping registration procedure using binary-encoded light. (a) The initial projected features as observed by the camera. (b–h) The binary encoding of these features. Observing this series of patterns makes establishing correspondence between the camera and projector features very robust.

An example of such a binary encoding scheme is demonstrated in Figure 3.16. In this example, a projector displays a set of 10×10 features that have been uniformly sampled in the projector image. The camera initially observes the features and can determine if it has seen the correct number; however, there is no way of knowing the correspondence between the projected features and observed features (i.e., the projected image could be upside-down, on its side, etc.). When using the binary encoding scheme, each feature in the projector space is assigned a unique integer ID from 1 to M in scanline order. In our example, $M = 100$ because we have 100 features, and the IDs are assigned row by row. The total number of bits required to represent 100 is 8, requiring a total of 8 frames to be projected. Each frame is denoted by f_i, where $1 \leq i \leq 8$. In each f_i, only the features that have a 1 in their ith bit position are projected. Frame f_1 projects only the features whose IDs have 1 in their first (least significant) bit, f_2 projects only the features whose IDs have 1 in their second bit, and so on. For example, feature ID 51, corresponds to a circle of radius 5 centered at location $(5, 510)$ in the projector coordinate space. The ID of this feature is 51, binary encoded as 00110011. So, its corresponding feature will be projected only in the frames f_1, f_2, f_5, and f_6 and *not* in f_3, f_4, f_7, and f_8. This procedure requires synchronization between the projector and camera. With synchronization enabled, when the feature is observed in certain frames and not in others, its binary-encoded ID can be easily recovered. Figure 3.16 shows an example of recovering these binary-encoded IDs. A red box is placed around feature ID 51. Once the IDs are detected, the actual correspondences $(x_i, y_i) \rightarrow (u, v)$ are easily recovered by associating the known (x_i, y_i) in the projector space to the centroids of the observed features in the camera's coordinates. Once this correspondence is established, this data can be used either to solve for parametric approaches (such as homographies) or to construct a nonparametric mesh for warping.

3.3.3   Data Fitting for Parametric Functions

Given the correspondences between (x_i, y_i) and (u, v), parametric functions can be computed by setting up a system of linear equations and solving for the unknown parameters in a least-squares sense. First, we discuss how to compute the homography used in linear methods for parametric surfaces.

Because homographies are computed between many different coordinate frames (e.g., camera-to-camera, projector-to-camera, camera-to-display), we discuss this mapping in terms of a *domain* coordinate (S, T) to a *range* coordinate (s, t). The set of correspondences between these two is denoted by $(S, T) \rightarrow (s, t)$. The (S, T) and (s, t) are related by a homography defined by nine parameters $(h_1, h_2, \ldots, h_8, h_9)$ and given by

$$\begin{pmatrix} sw \\ tw \\ w \end{pmatrix} = \begin{pmatrix} h_1 & h_2 & h_3 \\ h_4 & h_5 & h_6 \\ h_7 & h_8 & h_9 \end{pmatrix} \begin{pmatrix} S \\ T \\ 1 \end{pmatrix},$$

thus,

$$\begin{aligned} sw &= Sh_1 + Th_2 + h_3, \\ tw &= Sh_4 + Th_5 + h_6, \\ w &= Sh_7 + Th_8 + h_9. \end{aligned} \quad (3.5)$$

From the mathematics of homogeneous coordinates, we know that

$$\begin{aligned} s &= \frac{sw}{w}, \\ t &= \frac{tw}{w}. \end{aligned} \quad (3.6)$$

Because the homography is defined up to an arbitrary scaling, h_9 can be set to 1. Replacing Equations (3.5) in Equations (3.6), we get the following two equations with respect to the eight unknowns $h_1, \ldots, h_8$:

$$\begin{aligned} Sh_1 + Th_2 + h_3 - sSh_7 - sTh_8 - s &= 0, \\ Sh_4 + Th_5 + h_6 - tSh_7 - tTh_8 - t &= 0. \end{aligned} \tag{3.7}$$

Given a set of N corresponding points $(S_i, T_i) \rightarrow (s_i, t_i)$, $0 \le i < N$, we can set up the following linear system from Equations (3.7):

$$Ah = b,$$

where

$$\begin{pmatrix} S_1 & T_1 & 1 & 0 & 0 & 0 & -s_1 S_1 & -s_1 T_1 \\ 0 & 0 & 0 & S_1 & T_1 & 1 & -t_1 S_1 & -t_1 T_1 \\ S_2 & T_2 & 1 & 0 & 0 & 0 & -s_2 S_2 & -s_2 T_2 \\ 0 & 0 & 0 & S_2 & T_2 & 1 & -t_2 S_2 & -t_2 T_2 \\ \vdots & \vdots & \vdots & \vdots & \vdots & \vdots & \vdots & \vdots \\ S_N & T_N & 1 & 0 & 0 & 0 & -s_N S_N & -s_N T_N \\ 0 & 0 & 0 & S_N & T_N & 1 & -t_N S_N & -t_N T_N \end{pmatrix} \begin{pmatrix} h_1 \\ h_2 \\ h_3 \\ h_4 \\ h_5 \\ h_6 \\ h_7 \\ h_8 \end{pmatrix} = \begin{pmatrix} s_1 \\ t_1 \\ s_2 \\ t_2 \\ \vdots \\ s_N \\ t_N \end{pmatrix}.$$

A minimum of four point correspondences are necessary to set up the above equations. Solving for the parameters of vector h can be done in a variety of ways, such as LU, QR, or SVD matrix decomposition. When more than four correspondences are used, the matrix A has the same number of columns as the number of unknowns (eight, in this case), and the number of rows is two times the number of correspondences N. Thus, in this case, A is a $2N \times 8$ matrix. The goal is to find an h that minimizes the Euclidean norm $\|Ah - b\|_2$, thus providing an optimal h in a least-squares sense. This is solved by computing $(A^T A)x = A^T h$ using the pseudoinverse $A^T A$ such that $h = (A^T A)^{-1} A^T b$. Again, any matrix solver can be used to compute h. Due to small errors in localizing the point correspondences between devices, especially camera and projector mapping, we strongly recommend using more than four point correspondences when computing homographies between devices.

Figure 3.17 shows an example of geometric correction using a homography for a single projector. Figure 3.17(a) shows the image of 20×20 features that is projected by the projector. Figure 3.17(b) shows the camera image of these features, followed by the inverse of the computed homography in Figure 3.17(c). This homography is used to map camera coordinates to the projector coordinates. In this case, we assume that the image co-

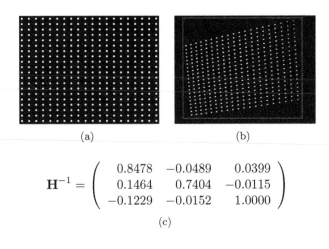

(a)   (b)

$$\mathbf{H}^{-1} = \begin{pmatrix} 0.8478 & -0.0489 & 0.0399 \\ 0.1464 & 0.7404 & -0.0115 \\ -0.1229 & -0.0152 & 1.0000 \end{pmatrix}$$

(c)

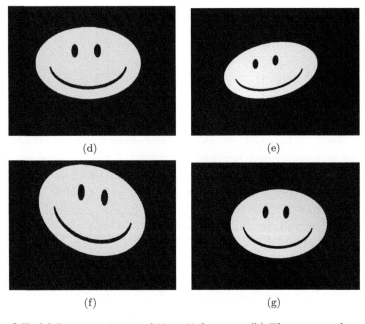

(d)   (e)

(f)   (g)

Figure 3.17. (a) Projector image of 20×20 features. (b) The camera observation of these features. (c) The homography, $\mathbf{H}^{-1}$ computed between these features and the mapping to the normalized projector coordinates ranging between $[0, 1]$ in the vertical and horizontal direction. (d) The original image $I(s, t)$. (e) This original image in (d), when projected from the projector and captured by the camera, looks distorted. (f) Image warped to the projector's coordinate frame using $\mathbf{H}^{-1}$. This is the image that should be projected by the projector to achieve geometric correction. (g) This warped image in (f), when projected by the projector and captured by the camera, looks geometrically correct.

ordinates are coincident with the camera coordinates, as is assumed for nonparametric displays. The camera coordinates within the red bounding box are normalized to the range $[0, 1]$ and mapped to the normalized image coordinates (s, t) of the image $I(s, t)$. Figure 3.17(d) shows the original image $I(s, t)$, which, when projected by the projector, looks distorted from the camera's point of view, as shown in Figure 3.17(e). Figure 3.17(f) shows the same image after applying the warp $\mathbf{H}^{-1}$. Note that this image looks rectified and, when projected from the projector and captured by the camera, as shown in Figure 3.17(g), looks geometrically correct.

In a similar manner, the parameters for nonlinear parametric equations, as in Equations (3.4), can also be solved. The only difference in this case is that the linear system of equations creates a matrix A of dimension $2N \times 20$ instead of $2N \times 8$. This is due to the greater number of unknown parameters in a nonlinear equation as $a_i, b_i, 1 \leq i \leq 10$, in Equations (3.4). Hence, in this case, at least 10 correspondences are required to solve for the a_i and b_i, but we recommend a larger number of correspondences for greater accuracy.

3.3.4 Piecewise-Linear Methods

Piecewise-linear methods do not involve parameter computation. Instead, the geometric correction is performed completely based on a piecewise-linear mapping of the projector coordinates to camera coordinates. The tessellation is implemented using standard Delaunay triangulation of the 2D points. The geometric correction is implemented using bilinear interpolation of these mappings at the vertices of the triangles. Figure 3.18 shows an example. In this example, a single projector is projecting into the corner of two walls. The projector projects 100 features that are observed and detected in the camera as seen in Figure 3.18(a). A bounding box is used to establish the mapping from the camera to the image $I(s, t)$. The camera coordinates within this bounding box are normalized to the range $[0, 1]$ and mapped to the normalized image coordinates (s, t). The detected features in the camera coordinate space are then triangulated using Delaunay triangulation to form a tessellated mesh T_C. This mesh is automatically aligned with the image $I(s, t)$ as shown in Figure 3.18(b). This image in Figure 3.18(b), when input to the projector to project on the corner of two walls, results in a distorted image. This image as captured by the camera is shown in Figure 3.18(c). To correct for this, the corresponding features in the projector coordinate space are connected in exactly the same manner as in T_C to create the corresponding tessellated

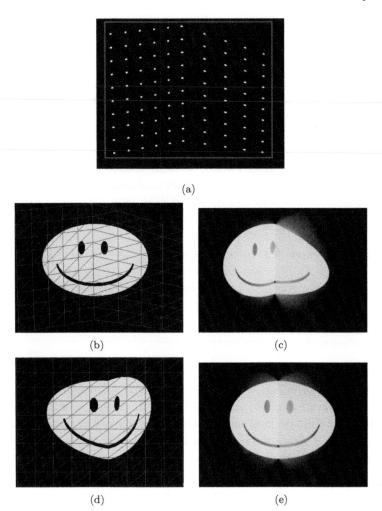

(a)

(b)   (c)

(d)   (e)

Figure 3.18. (a) Camera image of 100 projector features projecting into the corner of two walls. (b) The detected features in the camera and hence the image coordinate space $I(s,t)$ are tessellated to create the mesh T_C. (c) I, when projected by the projector on the wall and captured by the camera, looks distorted. (d) The corresponding mesh is created in the projector space by connecting the corresponding features in the same manner as T_C to create the tessellated mesh T_P. (e) The image warped from $I(s,t)$ shown in (c) is projected by the projector and then captured by the camera. It looks geometrically correct to a viewer at the camera location.

mesh in the projector space T_P. The pixels within a triangle in T_P are then picked from the corresponding triangle in T_C (and hence from $I(s,t)$) using bilinear interpolation. This warped image created from $I(s,t)$ by interpolating triangle-by-triangle to the corresponding tessellated mesh in the projectors coordinate space is shown in Figure 3.18(d). Figure 3.18(e) shows the image captured by the camera when this warped image in Figure 3.18(d) is projected by the projector. Note that even for an image projecting in the corner of the wall, it looks geometrically correct from the camera's point of view. This method can be used for either nonparametric or parametric surfaces.

Note that care must be taken in these nonparametric approaches to ensure that the projector's feature sampling matches the display surface. For example, in Figure 3.18(a), the features are able to capture the corner of the wall accurately. For display surfaces that are relatively smooth, this is not a problem. For display surfaces with sharp discontinuities, like the wall example in Figure 3.18(a), this can be a serious issue, requiring dense sampling or selective sampling.

3.4 Discussion

This chapter covered several methods to achieve geometric correctness for different display surface types. The various advantages and disadvantages of these different methods are discussed here along with other issues pertaining to geometric registration.

3.4.1 Planar Displays

Restricting the display surface to be planar has many benefits. The most notable is the ability to specify the corrective warp as a compact parametric function. Although point correspondences between the projector and camera are necessary to compute these parametric functions, once computed, the correspondence information is no longer needed.

The most compact form of this parametric function occurs when a 3×3 homography is used. This is especially conducive to mosiacing multiple cameras, which allows for scalable solutions as discussed. The significant drawback to this approach is its sensitivity to any nonlinearities in projectors and cameras. Because the entire approach is based on these perspective planar transforms, a strong assumption that the entire system is linear is being made. This means that the planar display is truly planar, and that the camera and projectors have no nonlinearities such as lens distortion. Any nonlinearities in any part of the system will effect the entire computation of geometric-correction homographies. In reality, very few devices, cameras and projectors both, behave in a purely linear fashion. Moreover, the walls in our buildings and rooms are not as planar as we might assume. Thus, small pixel errors are often hard to avoid. As a result, this approach often requires special display surfaces that have high-tension springs to pull the display material very tight to ensure planarity. Finally, while projector radial distortion is typically minimal, the effect can be compounded as more and more projectors are added together. This is particularly troublesome for the multiple-camera approach that may use several concatenated homographies to align the projector. As previously mentioned, Yang et al. [102] show that the zoom setting of some projectors affects the radial distortion enough to introduce a few pixel errors in homography-based approaches. This means that the projector's usable zoom range has to be fixed to the positions that minimize radial distortion, typically the default zoom setting. Thus, while this approach is mathematically sound, in practice it can raise issues that must be seriously considered, in particular, the costs of truly planar surfaces and the possibility of restricting projector optical settings.

The nonlinear parametric method offers a reasonable solution to this problem and can handle small projector nonlinearities. However, since lens distortions often have higher-order nonlinearities, especially for short-throw projectors, a cubic function is not sufficient to correct for all such nonlinearities. Further, since a cubic polynomial function is not perspective-projection invariant, this method assumes that the projectors are aligned in a near-rectangular fashion, i.e., with little keystoning. This assumption restricts the use of awkwardly oriented projectors. Thus, they work well for rear-projection systems where projectors are on a rack behind the screen where it is easy to have them almost perpendicular to the screen. However, for front-projection systems, considerable perspective projection leads to large keystone distortions. This is due to mounting of projectors on the ceiling from where they are tilted to project onto a screen on the ground. In such cases, this method may not yield an accurate alignment.

3.4.2   Nonplanar Display Surfaces

Parametric nonplanar surfaces. The approach for parameterized curved surfaces is a nice extension of planar surfaces. This approach has the benefit of not requiring the projectors to be perfectly linear. Small nonlinearity will be easily encoded in the mapping between the projectors and display surface. The only drawbacks are the physical construction of a specialized cylindrical display surface that is parametric and the need to attach a checkerboard fiducial at the top and bottom of the surface.

Nonparametric surfaces. This approach uses the camera image to find the mapping of projector pixels and a viewing location. This direct mapping from the projector pixels to the viewing location encodes all distortion present in the path of the projected image to the viewer's eye, including nonlinearities in the projector's image (such as radial distortion) as well as distortion due to the nonplanar display. And while we motivate this approach using nonplanar surfaces, there is no reason it cannot be applied to planar surfaces, especially if the projectors are exhibiting noticeable radial distortion.

It should be noted, however, that this approach does assume that the camera used for the geometric registration exhibits little or no distortion, since the camera image provides an interchangeable mapping between $C(u, v)$ and $I(s, t)$. This means that mistakes in the camera's position (e.g., a slight rotation of the camera about the optical axis) will be embedded in the mapping between the camera image and desired image and can cause noticeable artifacts when the desired image is warped and then projected.

The main drawback of this approach is scalability. Unlike the planar display set-up, it is not as easy, and often ill-advised, to construct mosiac images from multiple cameras observing a nonplanar surface. Instead, for large-scale arrays, this approach requires a very wide field of view, such as a camera with a fisheye lens. This lack of mosaicing ability typically restricts these approaches to relying on a single camera to observe the entire projector array. This leads to the problem of resolution mismatch between the camera and the projectors, which can make it difficult to establish the $P(x_i, y_i) \rightarrow C(u, v)$ mapping unless very large features are projected. Ever-increasing quality in commodity camera resolution is helping to combat this problem.

3.4.3 Camera Distortion

All of the techniques presented here assume that the camera exhibits no nonlinear distortion (i.e., no lens distortion) and therefore acts as a perfect linear pinhole camera. When the camera does exhibit noticeable distortion, several problems arise. For example, homography estimations will be incorrect due to errors introduced by the camera nonlinearities. This is compounded when multiple homographies are concatenated, such as with the multiple-camera approaches. For nonparametric approaches, nonlinearities in the camera manifest in geometrically registered imagery that looks radially distorted.

One solution to this problem is to pre-calibrate the camera to remove radial distortion. This requires each camera to apply a lens-distortion correction (see Appendix C).

3.4.4 Head Tracking

For displays where head tracking is desired, the parametric approaches are more advantageous because the image generated on the display is known to map directly to the display surface in a wallpaper fashion. For such environments, the rendering subsystem can generate images that appear correct for the viewer's head location and display this on the display surface.

For the nonparametric approaches, the mapping from the camera to the image that creates the correct image on the display is tied to a particular viewing location. Updating the image to be displayed based on a new head-track location can create visible inconsistencies and distortions. Thus, nonparametric techniques are not appropriate for the head-tracking scenario.

3.5 Summary

This chapter discussed image-based warping approaches to make geometrically correct imagery using a collection of casually positioned projectors for both parametric and nonparametric display surfaces. Camera-based geometric-registration techniques were described that use one or more cameras to observe the projected imagery in order to compute the parameters to guide the image-based correction for particular set-ups and viewing locations. Note that the geometric-registration techniques discussed in this chapter ensure geometric correctness only. Effects from nonuniform illumination and intensity balancing across the projectors is the subject of Chapter 4. In addition, implementation details on image generation for these multi-projector displays are described in Chapter 5.

4

Color Seamlessness

THE COLOR OF MULTI-PROJECTOR displays often shows significant spatial variation that can easily break the illusion of having a single display. The variation can be caused by device-dependent issues, such as *intra-projector color variation* (color variation within a single projector) and *inter-projector color variation* (color variation across different projectors), or by device-independent issues such as non-Lambertian or curved display surfaces, inter-reflection, and so on. Further, deliberate overlapping of projectors introduces significant brightness variation in the overlapped regions. In this chapter, the problems of color variation are first analyzed in detail. Next, the state-of-the-art solutions to address the color-variation problem and to achieve a seamless display are discussed.

To aid in understanding the causes of the color-variation problem, it is useful to have some background knowledge on color, displays, and human perception. A brief introduction to these topics is given in the following sections. A more detailed exposition is available in Appendixes A and B.

4.1 Background

Color is defined as the response created in a sensor, such as an eye or a camera, by a spectrum of light S. The spectrum S is a function of wavelength λ, and $S(\lambda)$ gives the amount of light of wavelength λ present in the spectrum. The range of wavelengths from 400 nm to 700 nm are visible to the human eye, as shown in Figure 4.1. We are only concerned with spectra of light that are visible by the human eye; therefore, $S(\lambda)$ will be defined within the visible range of wavelengths.

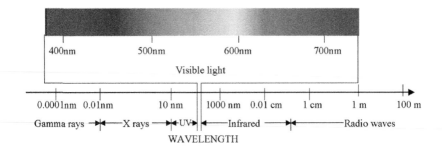

Figure 4.1. The visible spectrum of light.

Spectra can be of different types, as shown in Figure 4.2(a). A spectrum consisting of light of only one wavelength indicates a *monochromatic* color. A laser pointer provides monochromatic color; however, this kind of spectrum is rare in nature. A spectrum that contains equal amounts (relative power) of all different visible wavelengths indicates an *achromatic* color. Humans associate a hue or chroma with every color (such as red, blue, or green). Since achromatic colors have equal amounts of all wavelengths, they produce shades of gray and do not create any sensation of hue, hence the name achromatic. The most common spectra in nature have different amounts of different wavelengths and indicate *polychromatic* colors.

We first examine how a polychromatic spectrum relates to our perception of a color since that is the most common spectrum we encounter. We define a few terms that describe some common properties associated with

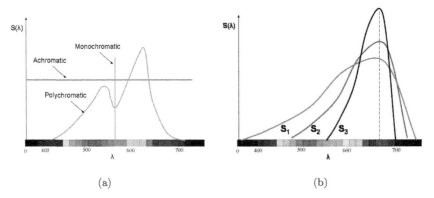

Figure 4.2. (a) Different types of spectra. (b) Properties of a spectrum.

a color: brightness/intensity, luminance, hue, and saturation. The hue and saturation are together called the chrominance of a color. The *intensity* of the color is the amount of light energy contained in the spectrum of the color and can be estimated by the area under the spectrum. The *chrominance* gives us the sense of colorfulness. The *hue* is defined by a dominant wavelength, which can be thought of as the weighted mean of all the wavelengths weighted by their relative amount in the spectrum. This means that the color produces a sensation of hue similar to the hue of its dominant wavelength. The *saturation* is the entity that gives a color its vibrancy and is the amount of achromatic color present in a color. This can be thought of as the variance of the spectrum from the dominant wavelength. For example, pink has a dominant wavelength of red, but it is an unsaturated version of red since white is combined with red to get pink. In Figure 4.2(b), we illustrate these properties. We show three spectra, S_1, S_2, and S_3. Each spectrum is drawn approximately in the color we see when we perceive them. Note that all of these spectra have the same hue but differ in saturation. S_3 is the most saturated, and S_1 is the least saturated. Further, as is evident from the area covered by the spectra, they have different values of intensity/brightness. Another important property of color is *luminance*, which is defined as the *perceived* brightness of a color. Luminance depends on the sensitivity of the human eye and is different from intensity.

Colors can be defined in a three-dimensional space (see Appendix A for details on how to arrive at this space). The most common color space used for this purpose is a device-independent color space called the CIE XYZ color space, spanned by three orthogonal axes X, Y, and Z, as shown in Figure 4.3(a). The visible colors only occupy a subset of the 3D space shown by the orange conical volume. The Y-axis corresponds to the luminance of a color. The chrominance can be derived from the three independent axes X, Y, and Z. It is represented by chromaticity coordinates (x, y) defined by

$$x = \frac{X}{X + Y + Z} \quad \text{and} \quad y = \frac{Y}{X + Y + Z}.$$

An alternate representation of a color is a tuple of its luminance and chrominance, (Y, x, y).

The significance of the chromaticity coordinates is important to understand. Note that (x, y) is generated by projecting a color in the 3D space onto a 2D plane given by $X + Y + Z = \alpha$. Consider two colors $C_1 = (Y_1, x_1, y_1)$ and $C_2 = (Y_2, x_2, y_2)$ with the same chrominance but

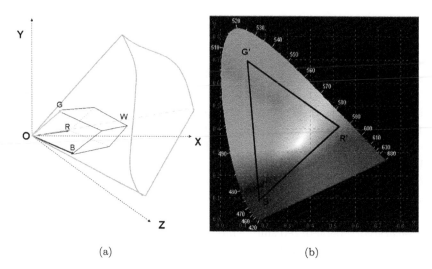

(a)   (b)

Figure 4.3. (a) The 3D XYZ color space. The orange conical shape shows the 3D volume spanned by all visible colors in this space. The rectangular parallelepiped spanned by the vectors OR, OG, and OB defines the 3D color gamut of a device. (b) The chromaticity diagram spanning a 2D space defining the chrominance of visible colors factoring out the luminance and the 2D color gamut of a device denoted by the black triangle given by the projection of R, G, and B on the plane $X + Y + Z = \alpha$.

different luminance, i.e., $(x_1, y_1) = (x_2, y_2)$. Simple mathematical computation reveals that the colors C_1 and C_2 lie on the same ray from the origin, i.e., their XYZ-coordinates are related by a scale factor k, $0 \leq k \leq 1$. This shows that all colors that lie on a vector starting at the origin in the XYZ space have the same chrominance but different luminance. The 2D projection onto the xy-plane essentially phases out the luminance, mapping colors with the same chrominance to the same chromaticity coordinates, ranging from 0.0 to 1.0. This 2D projection is called the *chromaticity diagram* and is illustrated in Figure 4.3(b). However, not all values of x and y represent visible colors. Only a subset enclosed by the horseshoe-shaped curve represent visible colors. This is essentially the projection of the 3D conical volume onto the xy-plane. Figure 4.3(b) shows the chromaticity diagram and the arrangements of the different hues in both saturated and unsaturated form. An achromatic hue is generated when $X = Y = Z$, i.e., $x = y = 0.33$. Thus, the point $W = (0.33, 0.33)$ is called the white point on the chromaticity diagram. All of the monochromatic colors in the visible spectrum, also called the spectral colors, form the boundary curve of the horseshoe region. The arrangement of the different wavelengths on this boundary is illustrated in Figure 4.3(b). The line joining two ends of the curve represents purples, which are not spectral colors. Our visible spectrum does not form a closed loop, but the boundary curve of the horseshoe region does. The purples can be thought of as the colors completing the loop. Thus, this boundary is called the non-spectral boundary of the chromaticity diagram.

Adding colors in the XYZ space is simply adding their coordinates in this space. For example, if a color $C_n = (X_n, Y_n, Z_n)$ is generated by adding two colors $C_1 = (X_1, Y_1, Z_1)$ and $C_2 = (X_2, Y_2, Z_2)$, then $C_n = (X_1 + X_2, Y_1 + Y_2, Z_1 + Z_2)$. Let us consider how the chrominance of these two colors are related. It can be shown that the chromaticity coordinate of C_n is given by

$$(x_n, y_n) = (a_1 x_1 + a_2 x_2, a_1 y_1 + a_2 y_2), \qquad (4.1)$$

where

$$a_1 = \frac{X_1 + Y_1 + Z_1}{X_1 + X_2 + Y_1 + Y_2 + Z_1 + Z_2}, \quad a_2 = \frac{X_2 + Y_2 + Z_2}{X_1 + X_2 + Y_1 + Y_2 + Z_1 + Z_2}.$$

The coefficients a_1 and a_2 have the following properties: $0 \le a_1, a_2 \le 1$ and $a_1 + a_2 = 1$. Thus, the chromaticity coordinate of the combined color is an affine linear combination of the chromaticity coordinates of the combining colors C_1 and C_2. This implies that the chromaticity coordinate of C_n lies on the straight line connecting the chromaticity coordinates of C_1 and C_2. This, in turn, shows that when two colors are added, the luminance of the resulting color is the addition of the luminance of the comprising colors and that its chromaticity coordinate lies on the straight line connecting the chromaticity coordinates of the comprising colors.

In an additive device such as a projector, colors are formed by superposition or additive mixture of different amounts of three primaries: red, green, and blue. An example of three such primaries is shown by the three points R, G, and B in the XYZ space in Figure 4.3(a). The corresponding device inputs that produce these R, G, and B primaries are given by $(1, 0, 0)$, $(0, 1, 0)$, and $(0, 0, 1)$, respectively. The XYZ-coordinate of black is usually $(0, 0, 0)$ and is the origin O in the XYZ space. All of the colors that can be reproduced by the device are defined by the rectangular parallelepiped spanned by the vectors OR, OG, and OB and is called the *3D color gamut* of the device. The XYZ-coordinate of the white W produced by the addition of the maximum input from each channel, i.e., input of (1,1,1), is given by the vector addition of the XYZ-coordinates of R, G, and B. The XYZ-coordinates of the gray colors produced by equal inputs from all three channels are located on the vector OW. Let R', G', and B' be the projection of R, G, and B on the chromaticity diagram in Figure 4.3(b). The triangle formed by these three points defines all the different chrominance values that can be reproduced by the device. This is called the *2D color gamut* of the device and is independent of the luminance. This is more commonly referred to as just color gamut. When specifying the 3D color gamut, the term 3D is explicitly mentioned. The 2D gamut is always a subset of the CIE chromaticity space. Also, notice that the position of the three primaries in the chromaticity diagram is away from the boundary of the visible spectrum, which says that they are not monochromatic light. Each primary is produced by a polychromatic light that spans a band of wavelengths and, as a result, produces a color that is less saturated than a monochromatic color. However, the primaries must be chosen such that they are *independent*, meaning that none of the primaries can be reproduced by a combination of two other primaries. In other words, the three primaries cannot lie on a straight line in the chromaticity diagram.

It is desirable to have a display where, given the properties of the primaries, one can predict, using simple formulae, the properties of any color produced by the combination of the primaries. Let us consider a display where the primaries R, G, and B are as follows. The luminance of R, G, and B are denoted by Y_R, Y_G, and Y_B, respectively. Their chromaticity coordinates are $R' = (x_R, y_R)$, $G' = (x_G, y_G)$, and $B' = (x_B, y_B)$, respectively. Let the channels of the display be denoted by $l \in \{r, g, b\}$. An ideal display should satisfy the following properties.

- Channel independence. The light projected from one channel is independent of the other two. This indicates that light from the other channels does not interfere with the light projected from another channel.

- Channel constancy. Only luminance changes with changing channel inputs. For example, for input $0.0 \leq r \leq 1.0$ of the red channel, the chromaticity coordinates (x_r, y_r) of r are constant at (x_R, y_R) and only the luminance Y_r changes. In other words, as the input varies from 0 to 1 for the red, green, and blue channels, the colors move from O to R, G, and B on the vectors OR, OG, and OB, respectively. However, the relationship between r and Y_r is usually nonlinear and is called the *input transfer function* (ITF). This ITF compensates for the fact that our eye has a nonlinear response to luminance.

- Spatial homogeneity. The response of all of the pixels of the display is identical for any input.

- Temporal stability. The response for any input at any pixel of the display does not change with time.

The property of optical superposition states that light falling at the same physical location from different sources adds up. The properties of channel constancy, channel independence, and optical superposition, along with the assumption that with an input of $(0,0,0)$ the display outputs zero light, indicate that the color projected at a pixel is a linear combination of the color projected by the maximum values of the red, green, and blue channels alone when the values of the other two channels are set to zero. Hence, for any input $c = (r, g, b)$, $0.0 \leq r, g, b \leq 1.0$, the luminance and chrominance of c are given by

$$Y_c = Y_r + Y_g + Y_b,$$

$$(x_c, y_c) = a_R(x_R, y_R) + a_G(x_G, y_G) + a_B(x_B, y_B),$$

where $0 \leq a_R, a_G, a_B \leq 1$ and $a_R + a_G + a_B = 1$. This follows directly from a similar calculation as is used to derive Equation (4.1). This property is referred to as the *linear-combination property*.

If a display satisfies linear combination, spatial homogeneity, and temporal stability, we can predict the color at all pixels at all inputs from the known response of the primaries at any one pixel of the display. Most traditional display devices such as CRT monitors satisfy these properties to a reasonable accuracy, or the deviation from this ideal behavior is simple enough to be modeled by simple linear mathematical functions [10]. However, as we will see in this chapter, a projector is not such an ideal device.

4.2   Color Variation in Multi-Projector Displays

The color-variation problem in multi-projector displays can easily break the illusion of having a single large display, even if the image is perfectly aligned across projectors geometrically. This is especially true when using commodity short-throw projectors, which help in providing a more compact set-up but do not have expensive optics to correct for the severe intra-projector color variation as illustrated in Figure 4.4. The most evident of the intra-projector variations is the radial spatial fall-off in luminance. This

Figure 4.4. Even in the presence of perfect geometric registration, color variation in multi-projector displays can be the sole cause of breaking the illusion of
having a single large display. (From [58]. © 2005 ACM, Inc. Included here by
permission. Display at Argonne National Laboratory.)

can be accentuated in rear-projection systems by the nature of the display
screen. Some of these problems can also be alleviated to a small extent
by favorable projector configurations. For example, overlapping projectors
across their boundaries can alleviate the problem of inter-projector luminance variation to some extent by compensating the low luminance at the
fringes with additional light from adjacent projectors. Based on these observations, several innovative projector configurations have been suggested,
including arranging projectors in honeycomb or overlapping fashion. Some
of these arrangements have been investigated in detail [59], and it would
be interesting to study such arrangements more in the future. However,
this clearly indicates that tiling projectors in an abutting fashion not only
leads to a rigid system but also to a display that shows severe variation in
color. Even for overlapping-projector configurations, it is evident that the
color-variation problem in a single projector cannot be removed completely
without using cost-prohibitive optics which are impractical in commodity projectors. In fact, color variation across different projectors cannot
be resolved completely even by using expensive optics. As a result, the
color-variation problem needs to be analyzed and characterized to arrive
at solutions that are effective and efficient.

The work in [57] presents a simple parametric space to study the color
variation of multi-projector displays. The parameters are *space*, *time*, and
input. Given a fixed input and time, the nature of the change of color
over space characterizes the spatial color variation. Similarly, given the

same pixel location and input, the change in color with time defines temporal color-variation characteristics. Finally, for the same pixel location and time, the color response with changing input defines input-response characteristics. Analysis of the color-variation problem in the context of this parametric space shows that multi-projector displays are different from traditional displays in many ways, so much so that assumptions that can be safely made about other display devices cannot be made for multi-projector displays.

The different device-dependent parameters of a projector that can cause color variation in a multi-projector display are also studied in [57]. These include position, orientation, zoom, lamp age, and projector controls such as brightness, contrast, and white balance. *Analyzing the effects of changes* in these parameters on color variation provides insights into the possible reasons for these variations. These analyses lead to the *key observation* that the most significant cause of the spatial variation in the color of a multi-projector display is the variation in luminance, especially for tiled displays consisting of projectors of the same make and model that vary negligibly in chrominance. In the rest of this section, we provide a summary of this detailed analysis and the related observations.

4.2.1 Measurement

Before we delve deep into the study of color variation, we present the details of the measurement process, instruments, and other extraneous factors that may have an effect on the analysis. The graphs and charts presented here are from [57] and are samples of the results achieved from different kinds of projectors. Since similar results were achieved from different kinds of projectors consistently, these results are representative of general projector characteristics.[1]

Measuring devices. The goal of the process of measurement is to accurately find the luminance and the chrominance properties at different points of the tiled display. There are two options for the optical sensors that one might use for this purpose.

A *spectroradiometer* is an expensive precision instrument that can accurately measure any color projected by the projector in a laboratory-calibrated device-independent 3D color space. However, a spectroradiometer measures only one point at a time at a very slow rate of about 1–20

---

[1] As test projectors for experiments, [57] used multiple Sharp, NEC, Nview, Proxima, and Epson projectors (both LCD and DLP) (at least 3–4 projectors of each model) and both front- and rear-projection systems.

seconds per measurement. Furthermore, it is difficult to measure the response of every pixel separately at a reasonable geometric accuracy. Thus, it is unsuitable for acquiring high resolution data.

A color camera, on the other hand, is relatively inexpensive and is suitable for acquiring high-resolution data in a very short time. There are many existing algorithms to find the geometric correspondence between the camera coordinates and the projector coordinates, assuring geometric accuracy of the data acquired. However, a color camera's limitation is its relative photometric inaccuracy as compared to a spectroradiometer.

All of the colors projected by the projector cannot be measured if the 3D color space of the camera does not contain the space of the projector, thus restricting us to work in a device-dependent 3D color space. Hence, cameras cannot be used for any chrominance measurements, only luminance measurements.

For the sake of accuracy, in [57], for point measurements, a precision spectroradiometer was used,[2] but for finding the spatial color/luminance variation across a projector, where color needs to be measured at potentially every pixel of a projector, a high-resolution digital camera was used. To reduce the photometric inaccuracies of the camera by a reasonable amount, the following considerations were taken into account.

- The nonlinearity of the camera was estimated using the algorithm presented in [25]. Every picture from the camera was linearized using these color look-up tables.

- It is important that the camera does not introduce more spatial variation than is already present on the display wall, i.e., the camera must produce flat fields when it is measuring a flat color. Most cameras satisfy this property at narrow aperture settings, especially below f/8, so the camera was set to such a narrow aperture.

- To ensure that a camera image is not under- or overexposed, simple under- or oversaturation tests were performed. The differing exposures were accounted for by appropriate linear scaling factors [25].

- To ensure geometric accuracy of the measurements, a geometric-calibration method was used to find accurate camera-to-projector correspondence. Any standard geometric-calibration mechanism can be used for this purpose [20, 38, 80].

---

[2]Photo Research PR 715.

Screen material and view dependency. For experiments on front-projection systems, [57] uses a close-to-Lambertian screen that does not amplify the color variations. Rear-projection screens are usually non-Lambertian, where measuring devices are sensitive to viewing angles. So, the spectroradiometer, was oriented perpendicular to the point being measured. For the camera, the view dependency cannot be eliminated. However, the camera was used for qualitative analysis of the nature of the luminance variation and not for generating an accurate quantitative mode. Hence, the view dependency was not critical.

Ambient light. Ambient light seen by the sensors was reduced as much as possible by taking the readings in a dark room, turning off all lights. When taking the measurement of a projector, all adjacent projectors were turned off. Further, black material was used to cover up the white walls of the room to avoid inter-reflected light.

4.2.2   Intra-Projector Variation

Intra-projector variation is the variation of color within a single projector. This was studied in detail with respect to the parameter space defined by input, space, and time.

Input variation. If a display satisfies channel independence and channel constancy, it should produce zero light for input of $(0, 0, 0)$, i.e., black. However, in projectors, because of light leakage, some light is projected even for black. This is called the *black offset*. This black offset can be up to 2% of the maximum luminance projected per channel but can deviate considerably across different projectors with a standard deviation of as large as 25%, even for projectors of the same make.

The contrast, or dynamic range, of each channel is defined as the ratio of the maximum luminance to the minimum luminance that can be projected from that channel. In a display where the black is close to zero, the contrast is close to infinity. In projectors, due to the presence of the black offset, the contrast for each channel and hence the display is adversely affected. Further, due to the presence of the black offset, the chromaticity for any channel at zero is the chromaticity of this achromatic black. As the inputs increase, the chromaticity reaches a constant value as it should for a device with channel constancy (see Figure 4.5). The contours shown in Figure 4.6 show how the gamut starts out as a single point for 0 in all three channels and then attains the final red triangle at the highest input

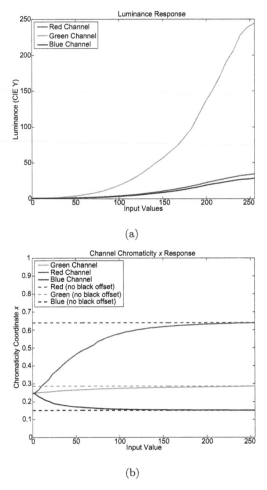

Figure 4.5. (a) Luminance response of the three channels. (b) Chromaticity x for the three channels. The shape of the curves for chromaticity y are similar. The dotted lines show the chromaticity coordinates when the black offset was removed from all the readings. (© 2003 IEEE. Reprinted, with permission, from [57].)

value. However, if this black offset is subtracted from the response of all inputs, the chromaticity-coordinate curves for the three channels (shown with dotted lines in Figure 4.5(b)) is constant [91, 90].

The black offset is characteristic of projectors with light-blocking technologies such as liquid crystal displays (LCD) or digital micromirror devices (DMD). Projectors using cathode-ray-tube technology do not show the black-offset problem to a perceptible extent.

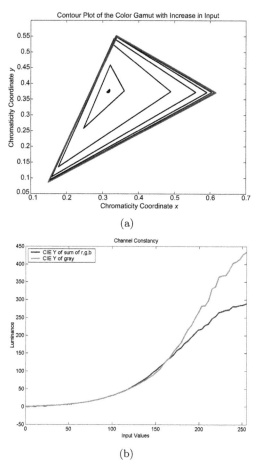

(a)

(b)

Figure 4.6. (a) Gamut contour as the input changes from 0 to 255 at intervals of 32. (b) Channel constancy of DLP projectors with white filter. (© 2003 IEEE. Reprinted, with permission, from [57].)

If the black offset is accounted for by a linear-offset term, almost all projectors exhibit the linear-combination property. However, some DLP projectors have a four-segment color wheel where a fourth clear filter element is used to enhance the contrast of grays beyond that achievable with a simple combination of red, green, and blue filters. Hence, the luminance of the grays is much higher than the luminance of the red, green, and blue put together, as illustrated in Figure 4.6(b). These devices cease to be three-primary devices and need to be handled differently. This is discussed in detail in Section 4.4.2.

In CRT monitors, the input transfer function (ITF) resembles a power function [5, 6, 10] usually called the "gamma" function. A power function assures a monotonic response. However, projectors typically have an S-shaped nonmonotonic response, as shown in Figures 4.5(a) and 4.10.

Spatial variation. Projectors are not spatially homogeneous. Figure 4.7 shows accurate luminance readings taken at equally spaced locations on

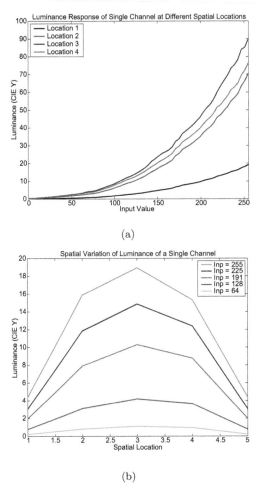

(a)

(b)

Figure 4.7. (a) Luminance response of the red channel plotted against input at four different spatial locations. (b) Luminance variation of different inputs of the red channel plotted against spatial location. The responses are similar for other channels. (© 2003 IEEE. Reprinted, with permission, from [57].)

the projector diagonal using the spectroradiometer. In Figure 4.7(b), the locations are named from 1 to 5 starting at the top-left corner position. The luminance reaches a peak at the center (location 3) and falls off at the fringes by a factor that may be as high as 80% of the peak luminance for rear-projection systems and about 50% for front-projection system. This considerable fall-off in luminance indicates that having wide overlaps between projectors in a multi-projector display can help us to get a better overall dynamic range.

However, the important thing to note here is that only the luminance changes spatially; the color gamut remains almost constant. Figure 4.8(a)

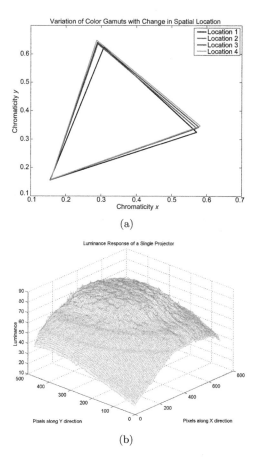

(a)

(b)

Figure 4.8. (a) Color gamut at four different spatial locations of the same projector. (b) Spatial variation in luminance of a single projector for input (1, 0, 0). (© 2003 IEEE. Reprinted, with permission, from [57].)

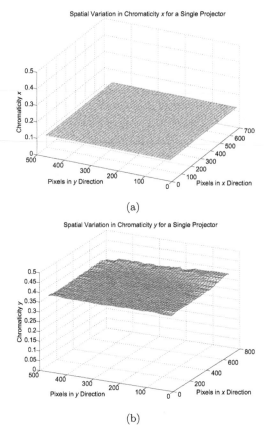

Figure 4.9. Spatial variation in (a) chromaticity coordinate x and (b) chromaticity coordinate y for maximum input of green in a single projector. (© 2003 IEEE. Reprinted, with permission, from [57].)

shows the gamut measured from the chromaticity coordinates of the primaries at their highest intensities at different spatial locations; the gamut varies minimally for the whole range of inputs. Figure 4.9 shows the spatial variation of the chromaticity coordinates of green to illustrate this. Note that both of the chromaticity coordinates x and y are spatially constant. Similarly, the normalized channel ITF is also spatially constant. This indicates that the shape of the luminance fall-off is similar for all inputs of a channel, with only a difference in range.

Given these observations from the spectroradiometer measurements, a camera was next used to measure the intra-projector spatial luminance variation at a much higher resolution, the luminance response shows a peak somewhere near the center and then falls off radially towards the fringes in an asymmetric fashion as shown in Figure 4.8(b). However, the shape of this fall-off varies from projector to projector due to other parameters, as will be discussed shortly.

The above observations can be easily explained. The chrominance depends on the physical red, green, and blue filters of the projectors that do

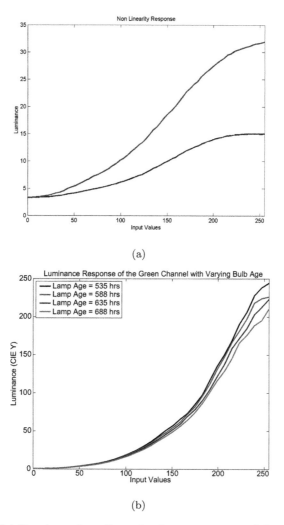

(a)

(b)

Figure 4.10. (a) Per-channel nonlinear luminance response of the red and blue channels. (b) Luminance response of the green channel at four different bulb ages. (© 2003 IEEE. Reprinted, with permission, from [57].)

not change spatially within a single projector. Hence, the chrominance is spatially constant. The luminance fall-off is due to the distance attenuation of light, further amplified by the non-Lambertian nature of the display. The asymmetry in the fall-off pattern gets pronounced with off-axis projection, as we will see in the following sections. This indicates that the orientation of the projector is responsible for this asymmetry.

Temporal variation. Finally, we find that the projectors are not temporally stable. The lamp in the projector ages with time and changes the color properties of the projector. Figure 4.10 shows a significant difference in luminance even within a short amount of time while the chrominance remains almost constant. The chrominance characteristics also drift a little after extensive use of about 800–900 hours.

4.2.3   Projector Parameters that Change Color Properties

Different projector parameters can change the color properties of a projector. This section presents the studies from [57] on the effects of varying these parameters on the color properties of a large-area multi-projector display and provides insights into the possible reasons for such effects.

Position. Position is defined as the distance of the projector from the wall along the axis of projection and the orientation, i.e., the alignment of the axis of projection with the display surface. Color properties were studied with two sets of experiments. In one, the orientation was constant while changing the distance from the wall, and in the other, the distance was constant while changing the orientation.

- Distance form the wall. Figure 4.11(a) shows the ITF as we move the projector at different positions along its axis of projection. The normalized ITF remains unchanged with this change in position. Further, the chrominance also remains constant. Finally, the shape of the spatial variance of the luminance also remains the same, as shown in Figures 4.7 and 4.8(b), but its absolute range changes.

  By moving the projector away from the wall, the projection area increases. Hence, the amount of light falling per unit area changes, but the nature of the fall-off does not change. The total energy remains the same, but the energy density changes with the distance from the center of the projector. Further, with the increase in distance, the

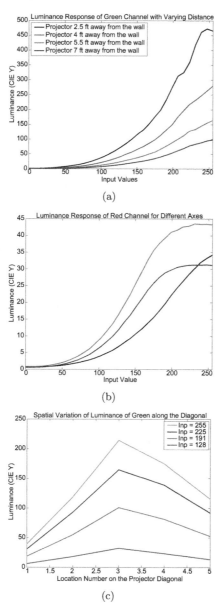

Figure 4.11. (a) Input transfer function (ITF) of the green channel as the distance from the wall is varied along the axis of projection. (b) Luminance response of the red channel with varying axis of projection. (c) Luminance response of different inputs in the green channel plotted against the spatial location along the projector diagonal for oblique axis of projection. (© 2003 IEEE. Reprinted, with permission, from [57].)

projector's throw ratio changes, which can cause a change only in luminance.

- Off-axis or orthogonal projection. In this set of experiments, the projector was at the same distance from the wall while it was rotated about the x-, y-, and z-axes to have an off-axis projection. In this case, the chrominance remains constant while the luminance response changes as shown in Figure 4.11(b) and (c). The nature of the spatial variation is no longer symmetric as in the case of orthogonal position (Figure 4.7). Near the longer boundary of the keystoned projection, which is physically farther from the projector, there is a larger luminance fall-off.

  As the orientation becomes more oblique, the luminance attenuation at the projector boundary farther from the screen increases, resulting in asymmetric fall-off. This is due to two reasons. First, the light from each pixel gets distributed over a larger area. Second, the angled surface receives less incident light due to the cosine effect. The results for vertical direction do not show a symmetry even for orthogonal projection due to offset projection. However, the chrominance remains constant, since moving the projector around does not change the internal filters of the projector and their set-up.

Projector controls. The projectors offer us various controls, such as zoom, luminance, contrast, and white balance. Knowing how these controls affect the luminance and chrominance properties of the projector can help us decide on the desirable settings for the projector controls that reduce variation within and across projectors. We can then select the best possible dynamic range and color resolution offered by the device; [57] performs experiments to study the effects of these controls on the intra-projector color variation.

- Zoom. Both luminance and chrominance were found to be constant with the change in zoom settings of the projector. With the change in zoom, the amount of light for each pixel gets distributed over a different area. For a focused projector, it is distributed over a small area, while for an unfocused projector, it is distributed over a larger area. However, the total area of projection remains the same, and the total amount of photon energy falling in that area remains constant. Hence, the light per unit area remains unchanged, while the percentage of light that each unit area receives from the different pixels changes.

- Brightness. The brightness control usually affects the ITF of the projector. Previous work mentions that usually the brightness control in displays changes the black offset [77]. However, in projectors, this control affects both the gain and the black offset of the ITF of *all three channels similarly and simultaneously*. The results are illustrated in Figure 4.12. As the brightness is increased, both the black offset and the gain of the ITF increase. However, if the brightness control is too low, the ITF gets clipped at the lower input range, and since the ITF remains at the same level for many lower inputs, the chromaticity coordinates also remain constant. At very high settings for the brightness control, some nonmonotonicity was observed in the ITF for the higher input range. As a consequence, the chromaticity coordinates also show some nonmonotonicity at the higher settings. Thus, it is ideal to have the brightness control set so that there is no clipping in the lower input range or nonmonotonicity at higher input ranges. For example, in these illustrations, the ideal setting is between 0.5 and 0.75.

- Contrast. Contrast also affects *all three channels similarly and simultaneously*. The results are illustrated in Figure 4.13. Previous work mentions that usually the contrast control changes the gain of the ITF [77]. The same was found to be true for projectors. As the gain increases, the luminance difference becomes significant enough at lower input ranges to push the chromaticity away from the gray chromaticity values towards the chromaticity coordinates of the respective primaries. However, the ITF starts to show severe nonmonotonicity at higher contrast settings, thus reducing the input range of monotonic behavior. So, the contrast setting should be in the monotonic range to maximally use the available color resolution.

- White balance. The white balance usually has a brightness and contrast control for two of the three channels, and the third channel acts as a reference and is fixed. The luminance and chrominance response changes in exactly the same way as for the independent brightness and contrast controls, but the change affects *only one channel at a time* instead of affecting all of them simultaneously. This way, the white balance controls the proportion of the contribution from each channel to a color, which in turn changes the white balance (Figure 4.14).

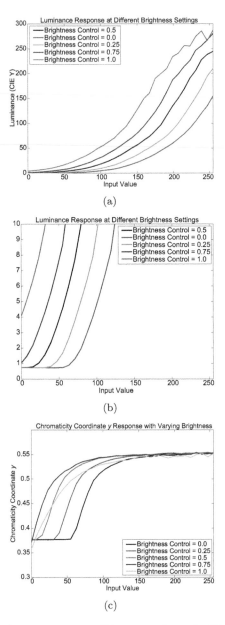

Figure 4.12. (a) Luminance response of the green channel with varying luminance settings. (b) Luminance response of the green channel with varying luminance settings zoomed near the lower input range to show the change in the black offset. (c) Chrominance response of the green channel with varying luminance settings. (© 2003 IEEE. Reprinted, with permission, from [57].)

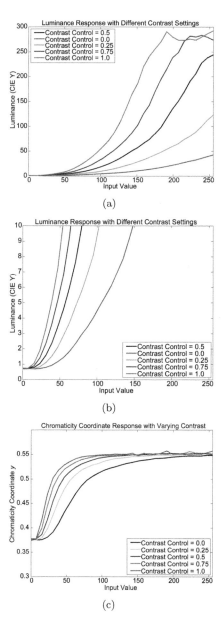

Figure 4.13. (a) Luminance response of the green channel with varying contrast settings. (b) Luminance response of the green channel with varying contrast settings zoomed near the lower luminance region to show that there is no change in the black offset. (c) Chrominance response of the green channel with varying contrast settings. (© 2003 IEEE. Reprinted, with permission, from [57].)

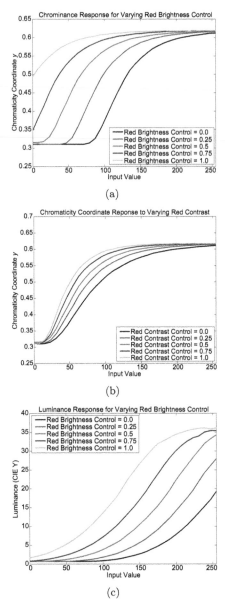

Figure 4.14. (a) Chrominance response of the green channel with varying green brightness settings for white balance. (b) Chrominance response of the red channel with varying red contrast settings for white balance. (c) Luminance response of the red channel with varying red brightness settings for white balance. (© 2003 IEEE. Reprinted, with permission, from [57].)

Projector Brand and Model	Red		Green		Blue	
	x	y	x	y	x	y
Sharp XG-E3000U	0.62	0.32	0.33	0.62	0.14	0.07
NEC MT-1035	0.55	0.31	0.35	0.57	0.15	0.09
nView D700Z	0.54	0.34	0.28	0.58	0.16	0.07
Epson 715c	0.64	0.35	0.30	0.67	0.15	0.05
Proxima DX1	0.62	0.37	0.33	0.55	0.15	0.07
Max Distance	0.085		0.086		0.028	

Table 4.1. Chromaticity coordinates of the primaries of different brands of projectors.

4.2.4 Inter-Projector Variation

Inter-projector variation is the variation of color properties across different projectors. Figures 4.15(a) and (b) shows the luminance and color-gamut response for the maximum intensity of a single channel for different projectors of the *same model* having exactly the same values for all the parameters defined in Section 4.2.3. There is nearly 66% variation in the luminance, while the variation in color gamut is relatively small. This small variation is due to physical limitations in the accuracy of manufacturing identical optical elements such as lenses, bulbs, and filters, even for the same brand of projectors.

Figure 4.18 also shows the near constancy of the high resolution chrominance response of a display wall made up of four overlapping projectors of the same model, projecting the same input at all pixels. Projectors of the same brand usually use the same brand of bulb (which have similar white points) and similar filters, which accounts for the similarity in the color gamut. However, this is not true for the grays of the DLP projectors that use the clear filter where the chrominance of grays differs significantly across different projectors due to large variation in the clear filter.

The difference in color gamut across projectors of different brands is greater than for the same model of projectors (Table 4.1). This is illustrated in Figure 4.15(c). However, this is relatively small when compared to the luminance variation.

4.2.5 Luminance Variation is the Primary Cause of Color Variation

The key observations from experiments and analysis of [57] presented in Sections 4.2.2, 4.2.3, and 4.2.4 is summarized as follows.

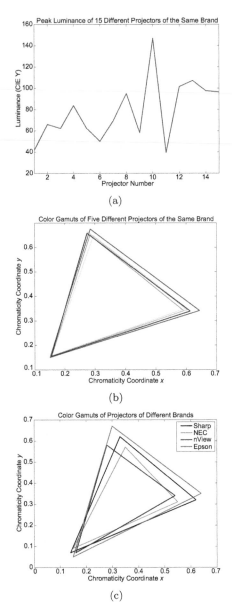

(a)

(b)

(c)

Figure 4.15. (a) Peak luminance of the green channel for 15 different projectors of the same model with the same control settings. (b) Color gamut of five different projectors of the same model. Compare the large variation in luminance in (a) with small variation in chrominance. (c) Color gamut of projectors of different models. (© 2003 IEEE. Reprinted, with permission, from [57].)

- Within a single projector's field of view, luminance varies significantly, while chrominance remains nearly constant.

- Luminance variation is dominant across projectors of the same model, but perceptually small chrominance variations also exist.

- The variation in chrominance across projectors of different models is small when compared to the variation in luminance.

- With the change in various projector parameters such as brightness, contrast, zoom, distance, and orientation, only luminance changes, while chrominance remains constant.

4.3   Modeling Color Variation

It was mentioned in Chapter 1 that building multi-projector displays presents many challenges, including data management, architecture, and geometric and color mismatches. Solutions to the data-management and driving-architecture problems in multi-projector systems [16, 40, 41, 42, 87] use existing parallel and streaming data models [65, 66, 69, 71, 73]. Solutions to the geometric-alignment problem [20, 24, 38, 39, 79, 80, 82, 83, 84, 103, 13] use the pinhole-camera model [29] to model individual projectors. However, there exists no model that describes the color variation in multi-projector displays that can be used to address the color-seamlessness problem.

Solutions for other issues related to multi-projector systems were borrowed from existing models for other types of systems and devices. For the color-variation problem, we cannot borrow models developed for other display devices, since, as we saw in the previous section, projectors differ from traditional display devices in a number of ways. For example, there exist models for colorimetry of single CRT monitors [5, 6, 10], but these models assume that for a given input, color remains spatially constant over the entire display. This assumption is not valid for projectors. Unlike CRTs, the physical space of a projector is decoupled from its display space, and several physical phenomena, namely distance attenuation of light, lens magnification, and non-Lambertian reflection of the display screen, cause severe spatial variation. For the same reason, projectors are also different from LCD panels. Further, there exists no model that describes color variations in multi-device displays with either CRT or LCD panels. Hence, [55, 58] use the lessons learned from the analysis in the previous section to design a color model for multi-projector displays.

4.3.1 Different Perspective on Colors in XYZ Space

Before we get deeper into the color-variation model and its description, we provide a new way of looking at colors in the XYZ space [55, 58]. The *intensity* L of a color is defined as the sum of its tristimulus values, i.e., $L = X + Y + Z$. This tristimulus simulation is used to indicate the total energy/power of the spectrum and provides an estimate of the absolute brightness of a color. The following lemmas define a few important properties of intensity. Although we prove these lemmas for only two colors, they can be extended easily to more colors.

Lemma 4.1. *The intensities of the colors that lie on a ray originating at the origin of the XYZ space have the same chromaticity coordinates, and their intensities are related by a scale factor.*

Proof: Let $C_1 = (X_1, Y_1, Z_1)$ and $C_2 = (kX_1, kY_1, kZ_1)$ denote two colors on the same ray originating at the origin in the XYZ space. The intensity of C_1 is $L_1 = X_1 + Y_1 + Z_1$. The intensity of C_2 is $L_2 = k(X_1 + Y_1 + Z_1) = kL_1$.

Lemma 4.2. *Addition of two colors results in a color whose intensity is the addition of the intensities of the comprising colors.*

Proof: Let a color $C_n = (X_n, Y_n, Z_n)$ be generated by adding two colors $C_1 = (X_1, Y_1, Z_1)$ and $C_2 = (X_2, Y_2, Z_2)$. Then the XYZ-coordinate of C_n is $(X_1 + X_2, Y_1 + Y_2, Z_1 + Z_2)$. The intensity L_n of C_n is given by

$$\begin{aligned} L_n &= X_1 + Y_1 + Z_1 + X_2 + Y_2 + Z_2 \\ &= L_1 + L_2. \end{aligned}$$

Lemma 4.3. *When two colors are added, the chromaticity coordinates of the constituent colors are added in proportion to their intensities to create the chromaticity coordinate of the new color.*

Proof: Equation (4.1) shows that when two colors $C_1 = (X_1, Y_1, Z_1)$ and $C_2 = (X_2, Y_2, Z_2)$ are added to create $C_n = (X_n, Y_n, Z_n)$, the chromaticity coordinate of C_n is given by an affine linear combination of the chromaticity coordinates of C_1 and C_2, where the linear coefficients are given by a_1 and

a_2 in Equation (4.1). Note that

$$a_1 = \frac{X_1 + Y_1 + Z_1}{X_1 + X_2 + Y_1 + Y_2 + Z_1 + Z_2}$$

$$= \frac{L_1}{L_1 + L_2} = \frac{L_1}{L_n}.$$

Similarly, $a_2 = \dfrac{L_2}{L_n}$. Hence, rewriting Equation (4.1), we get

$$x_n = (x_1, y_1)\frac{L_1}{L_n} + (x_2, y_2)\frac{L_2}{L_n}.$$

The above three lemmas along with the other properties of color presented in Section 4.1 provide the following important insights.

- From Section 4.1, we know that the luminance of colors lying on the same ray in the XYZ space are related by a scale factor k. Lemma 4.1 shows that the same property is valid for the intensity of a color.

- From Section 4.1, we know that when two colors are added, the luminance of the new colors is the sum of the luminance of the comprising colors. Lemma 4.2 shows that the same property is valid for the intensity of a color.

- From Section 4.1, we know that when two colors are added the chromaticity coordinate of the new color is an affine linear combination of the chromaticity coordinates of the comprising colors. Lemma 4.3 shows that the linear coefficients can now be easily defined in terms of the proportion of the intensities of the comprising colors.

Given the above properties of intensity, we see that all properties of luminance hold for intensity as well. In the rest of this chapter, only intensity (instead of luminance) and chrominance are used to define and handle color. This offers a better mathematical way to treat the color-variation model, primarily due to Lemma 4.3, which defines the chrominance as a linear combination of the ratios of intensity. Also, this allows the application of perceptual results on luminance to intensity when using such results in some color-calibration methods.

4.3.2 Notation

A color K at a point on the display is defined by the tuple (L, C), where L is the intensity defined by the sum of CIE tristimulus values of the color,

$X+Y+Z$, and C is the chrominance defined by the chromaticity coordinate (x, y):

$$x = \frac{X}{X+Y+Z}; \quad y = \frac{Y}{X+Y+Z}.$$

We use the notation $L = \text{int}(K)$ and $C = \text{chr}(K)$.

We introduce two operators for colors: the addition operator and the intensity-scaling operator. The addition operator is used when light from multiple sources superimposes in the overlap region.

- Addition operator. The addition of two colors $K_1 = (L_1, C_1)$ and $K_2 = (L_2, C_2)$ is defined as

$$K = (L, C) = K_1 \oplus K_2,$$

where

$$L = \sum_{i=1}^{2} L_i,$$

$$C = \frac{\sum_{i=1}^{2} L_i C_i}{L}.$$

By adding two colors, the intensity of the constituent colors add up, and the chrominance of the different colors are weighted by the intensity proportion of each of them (given by L_i/L). This follows directly from Lemmas 4.2 and 4.3. Note that $\oplus$ is both commutative and associative.

An important observation is made here. If $K_1 = (L_1, C_1)$, $K_2 = (L_2, C_2)$, and $C_1 = C_2$, then the chrominance C of $K = K_1 \oplus K_2$ is $C = C_1 = C_2$.

- Intensity-scaling operator. The intensity scaling of a color $K_1 = (L_1, C_1)$ by a scalar k is defined as

$$K = (L, C) = k \otimes K_1,$$

where

$$L = kL_1,$$

$$C = C_1.$$

Note that $\otimes$ does not change the chrominance of a color and that it distributes over $\oplus$.

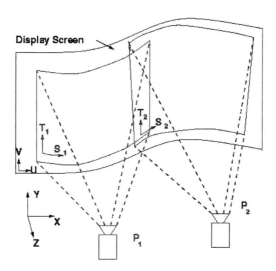

Figure 4.16. Projector and display coordinate space. (From [55]. © Eurographics Association and Blackwell Publishing Ltd 2005. Reproduced by kind permission of the Eurographics Association and Blackwell Publishing.)

4.3.3 Definitions

For a multi-projector display of N projectors, each projector is denoted as P_j, $1 \leq j \leq N$. The coordinates of the 2D display screen (not necessarily planar) are denoted by (u, v), and an individual projector's display region within this display screen is parameterized by (s, t). The geometry of the display screen S and the positions of the projectors (p_x, p_y, p_z) and the viewer $e = (e_x, e_y, e_z)$ are described in a 3D world coordinate system.

The display and projector coordinate pairs can be related by a *geometric warp G*:

$$(u, v) = G(s, t, p), \tag{4.2}$$

where $p = (p_x, p_y, p_z, \theta, \phi, f, S)$. The parameters f and (θ, ϕ) are the focal length and orientation of P, respectively, expressed in the world coordinate space. For all practical systems, p does not change because projectors and screen do not move relative to each other. Hence,

$$(u, v) = G(s, t). \tag{4.3}$$

Figure 4.16 shows a two-projector display wall. The blue and red quadrilaterals show the areas of projection of projectors P_1 and P_2 on the display screen.

A projector has three channels: r, g, and b. A channel is denoted by $l \in \{r, g, b\}$, and the corresponding input is denoted by $i_l \in \{i_r, i_g, i_b\}$, $0.0 \leq i_l \leq 1.0$.

The bidirectional reflectance distribution function (BRDF) for front-projection display or the bidirectional transfer distribution function (BTDF) for rear-projection display of the screen is dependent on the display coordinates (u, v) and the viewer location e and is denoted by $\Lambda(u, v, e)$. We assume that the BRDF/BTDF is independent of chrominance.

4.3.4   The Color-Variation Model

In this section, we introduce the function to model color in display devices in general and multi-projector display systems in particular [55, 58]. The function $E(u, v, i, e)$ that models the color characteristics of a display is defined as the color reaching the viewer e from the display coordinate (u, v) when the input to the device coordinates of all the display devices that contribute directly to the color at the display coordinate (u, v) is $i = (i_r, i_g, i_b)$. Note that more than one projector can contribute to the display coordinate (u, v) in the overlap region. Further, the function is defined only for devices with three primaries. A glossary of terms used in defining the color-variation model is presented in Table 4.2.

We derive $E(u, v, i, e)$ for a single projector display followed by the same for a multi-projector display.

Single-projector display. Let us consider one projector coordinate (s, t). Let $Q_l(s, t)$ be the maximum intensity that can be projected at that coor-

Notation	Parameter Description
l	$l \in \{r, g, b\}$ and denotes the red, green, and blue channels.
i_l	Input for channel l.
h_l	Spatially constant input transfer function of channel l, function of i_l.
c_l	Spatially varying chrominance of channel l, function of (s, t).
B	Spatially varying black offset, function of (s, t).
Q_l	Spatially varying maximum intensity function of channel l, function of (s, t).
N_p	Number of projectors overlapping at any pixel, function of (u, v).

Table 4.2.   Notation for the color-variation model parameters and their descriptions.

dinate from channel l. For any input i_l, the intensity projected is a fraction of $Q_l(s,t)$ and is given by $h_l(s,t,i_l)$, where $0.0 \le h_l(s,t,i_l) \le 1.0$. Let the chrominance projected at that coordinate for input i_l be $c_l(s,t,i_l)$. Thus, the color projected at coordinate (s,t) for input i_l in channel l is given by

$$
\begin{aligned}
D_l(s,t,i_l) &= (\ h_l(s,t,i_l)Q_l(s,t)\ ,\ c_l(s,t,i_l)\ ) \\
&= h_l(s,t,i_l) \otimes (\ Q_l(s,t)\ ,\ c_l(s,t,i_l)\ ).
\end{aligned}
$$

We call c_l the *chrominance function* and Q_l the *intensity function* of the channel l.

Let us first assume a projector with no black offset that follows both channel constancy and channel independency. Channel constancy results in $c_l(s,t,i_l)$ being independent of the channel input i_l. Hence,

$$
D_l(s,t,i_l) = h_l(s,t,i_l) \otimes (\ Q_l(s,t)\ ,\ c_l(s,t)\ ).
$$

The studies in Section 4.2 show that h_l is independent of (s,t). As a result, it is denoted by just $h_l(i_l)$ and becomes synonymous with the *input transfer function* (ITF) for channel l. We have

$$
D_l(s,t,i_l) = h_l(i_l) \otimes (\ Q_l(s,t)\ ,c_l(s,t)\ ). \tag{4.4}
$$

D_l is now expressed as a multiplication of two independent functions: h_l depends only on the input i_l, and (Q_l, c_l) depend only on the spatial coordinates (s,t). Figure 4.17(b), (a), and (d) illustrate the parameters Q_l, h_l, and c_l, respectively.

Channel independence indicates that the total color $T(s,t,i)$ projected at (s,t) for input $i = (i_r, i_g, i_b)$ is the superposition of colors from individual channels; that is,

$$
T(s,t,i) = D_r(s,t,i_r) \oplus D_g(s,t,i_g) \oplus D_b(s,t,i_b). \tag{4.5}
$$

In practice, we have to consider the black offset. This is represented by $(B(s,t), c_B(s,t))$, where $B(s,t)$ is the spatially varying intensity component called the *black intensity function* and $c_B(s,t)$ is the *black chrominance function*. Thus, the light projected by a projector is

$$
T(s,t,i) = D_r \oplus D_g \oplus D_b \oplus (B(s,t), c_B(s,t)).
$$

Figure 4.17(b) shows B when compared to the scale of Q_l. The spatial variation of B is better visible in its zoomed-in view in Figure 4.17(c).

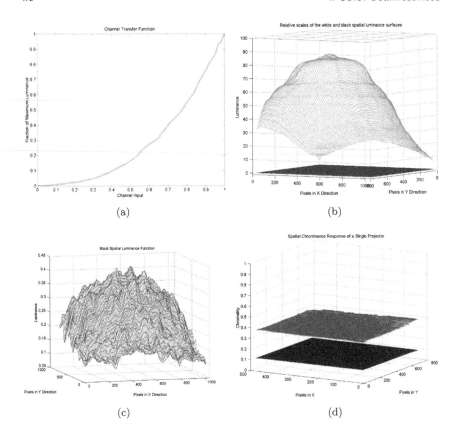

Figure 4.17. Reconstructing the color-variation model parameters for a single projector. (a) The input transfer function for the green channel. (b) The intensity function for the green channel and the black intensity function. (c) The zoomed-in view of the black intensity function. (d) The zoomed-in view of the chrominance function (x, y) for the green channel. (From [58]. © 2005 ACM, Inc. Included here by permission.)

Finally, for a viewer at e, the function $E(u, v, i, e)$ for a single projector is given by

$$E(u, v, i, e) = \Lambda(u, v, e) \otimes T(s, t, i), \qquad (4.6)$$

where $(u, v) = G(s, t)$.[3]

Multi-projector display. Using the color-variation model derived for a single-projector display, we can now derive the color-variation model for a multi-

---

[3]When p used in Equation (4.2) is not assumed to be static, G, Q_l, B, c_l, and c_B are dependent on p also.

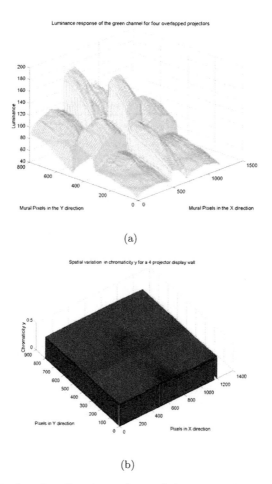

(a)

(b)

Figure 4.18. The function E at input $(1, 1, 0)$ for a tiled display made up of a 2×2 array of four projectors. The (a) intensity $\mathrm{int}(E)$ and (b) y-component of chrominance $\mathrm{chr}(E)$ are reconstructed from a camera image. ((a) from [58]. © ACM, Inc. Included here by permission. (b) from [55]. © Eurographics Association and Blackwell Publishing Ltd 2005. Reproduced by kind permission of the Eurographics Association and Blackwell Publishing.)

projector display. Let N_P denote the set of projectors overlapping at display coordinate (u, v). The function $E(u, v, i, e)$ for a tiled display is then given by the superposition of colors projected for the same input i from all the projectors overlapping at the display coordinate (u, v):

$$E(u, v, i, e) = \oplus_{j \in N_P} E_j(u, v, i, e). \tag{4.7}$$

For a Lambertian screen, whose image is not dependent on the viewer's location, Equations (4.6) and (4.7) are independent of e, and E is reduced to

$$E(u, v, i) = \oplus_{j \in N_P} T_j(s_j, t_j, i). \tag{4.8}$$

Figure 4.18 shows the function E for a four-projector display for input $(1, 1, 0)$.

4.3.5   Deriving Color-Variation Properties

The most important advantages of a color-variation model lie in the ability to derive different color-variation properties that can then be verified via experiments on physical set-ups. This, in turn, acts as a validation of the model itself. The color-variation model presented in Section 4.3.4 also results in the derivation and verification of some interesting color-variation properties. In this section, we state these properties, a summary of which is available in Table 4.3. However, we do not provide any mathematical proofs, for which the readers are directed to [55].

Intra-projector variation. The intra-projector *intensity* variation is modeled by the spatially varying intensity function Q_l for each channel l, as illustrated in Figure 4.17(b). Section 4.2 showed that this spatial variation in intensity is much more significant than the spatial variation in chrominance. However, some regions within a single projector can have chrominance vari-

Type of Variation	Intensity/ Chrominance	Parameters Responsible
intra-projector	intensity	spatial variation in Q_l
	chrominance	spatial variation in c_l; difference in shapes of $\bar{Q}_l$ across different channels
inter-projector	intensity	difference in h_l across projectors; difference in Q_l across projectors
	chrominance	difference in c_l across projectors; difference in proportion of M_l from different channels across projectors
overlap	intensity & chrominance	difference in $|N_p|$ at different spatial locations.

Table 4.3. Summary of the parameters responsible for different kinds of color variation in a multi-projector display.

ations in the form of visible color blotches. The first obvious reason for this intra-projector *chrominance* variation is the small spatial variations in the chrominance function c_l, as shown in Figure 4.17(d). However, the more interesting finding is that sometimes these color blotches result even if there is no spatial variation in c_l itself [55]. In fact, a chrominance variation can be entirely due to difference in the *shape* of the channel-intensity functions Q_l across different channels. The shape of Q_l is obtained by normalizing it. Note, however, that the *absolute* values of the intensity functions Q_r, Q_g, and Q_b can be different from each other, but this will not lead to chrominance variation.

Inter-projector variation. Difference in the intra-projector color variations across different projectors is called the inter-projector color variation. The inter-projector *intensity* variation is due to two factors: the input transfer function h_l that is not same for all projectors, and the intensity function Q_l that is different for different projectors. Even if Q_l for each projector is spatially constant at M_l, the differences in this constant itself will cause an inter-projector intensity variation. Essentially, each projector will vary in its overall intensity.

Similarly, even if we assume a spatially uniform c_l, inter-projector *chrominance* variation results due to differences in c_l across different projectors. More interestingly, even when there is no intra-projector spatial intensity/chrominance variation or inter-projector variation in c_l, inter-projector chrominance variation can still occur [55]. Let us assume Q_l for each projector to be spatially constant at M_l. In such a case, if the proportion of M_l coming from each channel differs from one projector to another, inter-projector chrominance variation will occur. This reflects the fact that a change in the intensity proportion of red, green, and blue across different projectors can trigger a change in chrominance. Even if the absolute values of M_l change across projectors, as long as the proportions are maintained, chrominance variation will not be seen. The differences in the absolute value of M_l would manifest in a variation in the overall projector intensity.

Overlap-region variation. The overlap-region variation defines the color difference between the overlap and non-overlap regions. The color in the overlap region is the superposition of the color from the individual projectors and is explicitly modeled by the addition operator in Equation 4.7. The cardinality of N_P ($|N_P|$) being different at every display coordinate causes this variation.

4.3.6 Relationship to Other Existing Display Color Models

A popular display color model that has been used extensively in color-matching and color-profiling applications for modeling various devices, such as cameras, scanners, printers, and monitors, assumes linear devices (i.e., $h_l(i_l) = i_l$) and hence uses a matrix that relates the RGB space of a device to the standard CIE XYZ space [30, 33, 92]. Note that in Figure 4.3(a), the transformation from the XYZ-coordinate to the parallelepiped spanned by the device primaries is essentially a linear coordinate transformation that can be represented by a 3×3 matrix whose rows are given by the XYZ-coordinates of the primaries at full intensity. Thus, the XYZ response of a color generated by input $i = (i_r, i_g, i_b)$ is given by

$$\begin{pmatrix} X_i \\ Y_i \\ Z_i \end{pmatrix} = \begin{pmatrix} X_r & X_g & X_b \\ Y_r & Y_g & Y_b \\ Z_r & Z_g & Z_b \end{pmatrix} \begin{pmatrix} i_r \\ i_g \\ i_b \end{pmatrix},$$

where (X_l, Y_l, Z_l) denotes the CIE XYZ response of the maximum intensity input of channel l. However, this model assume spatial constancy in color properties, and, hence, only one matrix is used to characterize the device. This works fine for the devices it is used for, since they usually do not show marked spatial variation in color. Since projectors do have marked spatial variation in color, an extension of this model was recently suggested for single projectors [68]. In this work, a spatially varying 3×3 matrix is assigned to each projector coordinate (s, t). This allows association of a different matrix with every pixel of the projector to account for the spatial variation. Further, to accommodate black offset, a 3×4 matrix was suggested [91, 90] as follows:

$$\begin{pmatrix} X_i \\ Y_i \\ Z_i \end{pmatrix} = \begin{pmatrix} X_r & X_g & X_b & X_B \\ Y_r & Y_g & Y_b & Y_B \\ Z_r & Z_g & Z_b & Z_B \end{pmatrix} \begin{pmatrix} i_r \\ i_g \\ i_b \\ 1 \end{pmatrix},$$

where (X_B, Y_B, Z_B) is the XYZ-coordinate of the black offset and also changes spatially.

However, this model does not handle multi-projector displays. For example, it is not defined how to construct these matrices for the pixels in the overlap region that come from more than one projector. Part of the restriction in defining the overlap regions in this model stem from the fact that a matrix does not differentiate the color properties in terms of our perception. For example, it is difficult to separate the terms that account for intensity and chrominance in the matrix.

The color-variation model presented in [55, 58] follows a different way to define the same information encoded in the matrix where the terms are clearly differentiated as intensity or chrominance. So, L and C for input i in this model are related to (X_i, Y_i, Z_i) as

$$(L_i, C_i) = \left(X_i + Y_i + Z_i, \left(\frac{X_i}{L_i}, \frac{Y_i}{L_i} \right) \right).$$

This separation in parameters helps in the classification and analysis of the intensity and chrominance properties differently. In essence, this color-variation model should be thought of as an abstract color model for imaging devices whose parameters should be evaluated using device-specific methods.

4.4 Color-Calibration Methods

The process employed to compensate for all of the different color variations to create a seamless multi-projector display is called *color calibration*. At the inception of tiled displays a couple of decades ago, exorbitant costs of projectors and monolithic rendering engines restricted the number of projectors in the display to a few (between two and four). These were set up in a rigid fashion so that the projectors did not overlap along their boundaries. In such a scenario, manual calibration was feasible during initial set-up or periodic recalibration. However, several man hours and many dollars were spent on color calibration using sophisticated and expensive optical elements. Even significant engineering feats have been tried to remove these variations, including using a common lamp to remove the color variations due to the projector bulb [72]. In this approach, the lamps of the multiple projectors were taken off and replaced with a common lamp of much higher power. Light was distributed from this common lamp to all the different projectors using optical fibers. Such manual methods are very cost-intensive and often demand extremely skilled labor.

Currently, projectors and rendering PCs are both commodity products. Thus, building a reasonably large multi-projector display (10–20 projectors) is quite affordable today. However, manual calibrations are no longer feasible, and of course not scalable. Therefore, automatic color-calibration methods were devised to create scalable displays. In this section, we present the different automated color-calibration techniques and analyze them [91, 90, 58, 56, 57, 101].

4.4.1   Color-Calibration Framework

Based on the color-variation model presented in the previous section, we can define a three-step framework for designing a color-calibration method.

1. Reconstruction. The function $E(u, v, i, e)$ defines the color that is seen by a viewer from the display. The first step of any color-correction algorithm should be to reconstruct the function $E(u, v, i, e)$ accurately. The accuracy and speed of the reconstruction depend on the sensors being used. The parameter e makes this function a view-dependent one. All existing color-calibration techniques assume a Lambertian screen and use $E(u, v, i)$ (as in Equation (4.8)) while reconstructing the function.

2. Modification. The reconstructed function shows perceivable color variation. Therefore, different parameters of $E(u, v, i)$ are modified to define the desired color response of the display that the algorithm wants to achieve. We call the new response generated the desired function, $E'(u, v, i)$. The function E' can be formally defined by a set of conditions it should satisfy defining the *goals* of the color-calibration method.

3. Reprojection. Usually, in a practical projector system, the user has no access to all of the various parameters that are modified to achieve E' (like Q_l, c_l, and h_l). To achieve E', only the input (i_r, i_g, i_b) to the different projector pixels is modified appropriately to get a color response identical to that defined by E'. Since this involves reprojection of modified inputs, this step is called reprojection.

Common goals. All of the color-calibration methods presented will be analyzed based on the above framework. An important component of this framework is the set of goals used while designing the desired function. This can be different based on the application and its demand on the accuracy of the color calibration. We define a few common goals that color-calibration algorithms implicitly or explicitly try to achieve.

- Strict color uniformity. The goal here is to make the response of two different projectors look the same. The desired function E' satisfies the property of *strict color uniformity* if the color (intensity and chrominance) of the light reaching the viewer from any two display coordinates for the same input is the same. Hence, $\forall i$, (u_1, v_1), (u_2, v_2),

$$|E'(u_1, v_1, i) - E'(u_2, v_2, i)| = 0.$$

  From Table 4.3, we note that the goal of strict color uniformity makes the parameters c_l, Q_l, and h_l identical at every display coordinate.

- Perceptual color uniformity. The goal of strict color uniformity is relaxed to generate the goal of perceptual color uniformity. Instead of the color reaching the viewer from two display coordinates being identical, this goal says that the color (intensity and chrominance) of light reaching the viewer from any two display coordinates can vary within a threshold δ that is imperceptible to the human eye. In other words, E' satisfies the property of *perceptual color uniformity* if $\forall i$, (u_1, v_1), (u_2, v_2),

$$|E'(u_1, v_1, i) - E'(u_2, v_2, i)| \leq \delta,$$

  where δ *cannot be detected by humans* and is a function that can depend on various parameters such as distance between (u_1, v_1) and (u_2, v_2), resolution of the display, distance of viewer, viewing angle, human perception limitations, and sometimes even the task to be accomplished by the user. From Table 4.3, we note that the goal of perceptual color uniformity makes the spatial variations in c_l and Q_l imperceptible in order to remove the intra-projector variation. Since h_l does not vary spatially within a single projector [57], it should vary imperceptibly across different projectors to address the inter-projector variation. Also, note that strict color uniformity is a special case of the perceptual color uniformity where $\delta = 0$.

- Strict photometric uniformity. Strict photometric uniformity is the result of relaxing the goal of strict color uniformity in a different way. Instead of achieving identical color (intensity and chrominance) at every display pixel, the goal here is to achieve just identical *intensity* at every display pixel without addressing the chrominance. In other words, E' satisfies the property of *strict photometric uniformity* if the intensity of light reaching the viewer from any two display coordinates for the same input is the same; that is, $\text{int}(E')$ is spatially constant. Hence, $\forall i$, (u_1, v_1), (u_2, v_2),

$$|\text{int}(E'(u_1, v_1, i)) - \text{int}(E'(u_2, v_2, i))| = 0. \qquad (4.9)$$

From Table 4.3, we note that the goal of strict photometric uniformity ignores the inter- and intra-projector *chrominance* variation and makes Q_l and h_l identical at every display coordinate.

- Perceptual photometric uniformity. The goal of strict photometric uniformity is further relaxed to define the goal of perceptual photometric uniformity, whereby the intensity at two display pixels need not be identical but can vary within a threshold imperceptible to the human eye. In other words, E' satisfies the property of *perceptual photometric uniformity* if the intensity of light reaching the viewer from any two display coordinates is within a threshold δ. Hence, $\forall i$, (u_1, v_1), (u_2, v_2),

$$|\text{int}(E'(u_1, v_1, i)) - \text{int}(E'(u_2, v_2, i))| \le \delta, \qquad (4.10)$$

where δ again is imperceptible to the human eye and can depend on several factors. From Table 4.3, we note that the goal of perceptual photometric uniformity makes the spatial variations in Q_l imperceptible in order to remove the intra-projector intensity variation. Also, as in the case of perceptual color uniformity, h_l should vary imperceptibly across different projectors to address the inter-projector variation. Finally, strict photometric uniformity is a special case of perceptual photometric uniformity with $\delta = 0$.

4.4.2   Gamut Matching

Gamut-matching methods try to address only the inter-projector variation using automated feedback from a low-resolution precision light-sensing device [91, 90, 101]. They assume that the intra-projector variation is negligible and that there is no overlap region. The correction is achieved in the following three steps and is illustrated for a simple two-projector case in Figure 4.19.

1. Reconstruction. An accurate light-measuring sensor, such as a spectroradiometer, is used to find the precise 3D color gamut and the input transfer function of each projector at one spatial location. The input transfer functions are used to linearize the inputs. These linearized inputs span a color gamut that is a rectangular parallelepiped

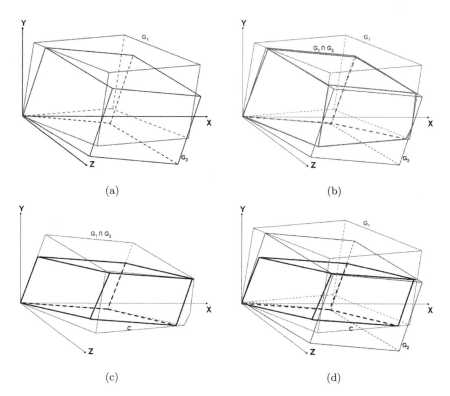

Figure 4.19. (a) The gamuts of two devices are shown in red and blue, respectively. (b) The intersection of these two gamuts is shown in green. Note that this is not a rectangular parallelepiped. (c) The largest rectangular parallelepiped contained within the intersection of two gamuts is the common gamut, shown in black. (d) The common gamut and the original gamuts of the two devices are shown together. This defines the function that will be used for each device to convert to the common gamut.

in the XYZ space. Thus, each projector i spans such a rectangular parallelepiped gamut called G_i. The gamuts of two such projectors are shown in Figure 4.19(a). As expected, there are colors in one gamut that cannot be reproduced by the other. As per the color-variation model of Section 4.3.6, there exists a matrix M_i that defines the linear transformation from the input in projector i (r_i, g_i, b_i) to the XYZ space. In matrix notation,

$$\begin{pmatrix} X \\ Y \\ Z \end{pmatrix} = M_i \begin{pmatrix} r_i \\ g_i \\ b_i \end{pmatrix}. \qquad (4.11)$$

2. **Modification.** Since the 3D color gamuts are different for different projectors, there are colors that can be produced by one projector and not by another. As a result, a *common 3D color gamut* needs to be defined that can be reproduced by all the different projectors. This is done in two steps.

   (a) The intersection of all the different color gamuts, $\cap_i G_i$, is identified. This is illustrated in Figure 4.19(b). Note that this intersection need not be a rectangular parallelepiped in shape. However, to assure that the common 3D color gamut shows the desired device properties of channel independency and constancy, the common 3D color gamut needs to be a rectangular parallelepiped, which effectively changes the set of primaries of the device but still keeps the relationship between the input space and the XYZ-coordinates linear.

   (b) The common 3D gamut is defined as the largest rectangular parallelepiped C contained in $\cap_i G_i$, as illustrated in Figure 4.19(c). In Figure 4.19(d), the same C is shown together with the original gamuts G_i. Note that since C is a subset of all of the G_i, all of the colors in C can be reproduced by all of the projectors. If (r, g, b) is designated as the input space for C, then we can find a matrix M that relates C to the XYZ space. Mathematically,

$$\begin{pmatrix} X \\ Y \\ Z \end{pmatrix} = M \begin{pmatrix} r \\ g \\ b \end{pmatrix}. \tag{4.12}$$

3. **Reprojection.** Now that the common 3D gamut is identified, the final step is to find a function F_i that defines a mapping from C to each gamut G_i. The input image is considered to be in the common gamut C, and a transformation defines the input in G_i that produces the same color as the input in C. In this case, F_i is different for each projector and is linear since all projectors are linear devices after the initial linearization step. It can be derived from Equations (4.11) and (4.12) that

$$\begin{pmatrix} r_i \\ g_i \\ b_i \end{pmatrix} = M_i^{-1} M \begin{pmatrix} r \\ g \\ b \end{pmatrix}.$$

This linear transformation is applied to the input of each projector to achieve the correction.

In terms of the framework presented in Section 4.4.1, this method reconstructs the function E, but only at a few display coordinates (u, v), usually at the center of each projector. The parameters c_l, Q_l, and h_l are estimated only at those sampled spatial locations. Under the assumption that these parameters do not change spatially within a single projector, the goal of this method is to achieve a *strict color uniformity* by making the Q_l and c_l identical for all projectors. Finally, this goal is achieved by modifying the inputs to the projectors.

Handling four-primary systems. The gamut-matching method described so far is only applicable to devices that use three primary colors. The three primary colors are linearly independent of each other and form a basis for describing all colors of a device; a color in the three-primary system can be represented by a *unique* combination of the three primaries. However, some DLP projectors commonly use a method they call "white enhancement" where they use a clear filter in addition to the red, green, and blue filters. This clear filter passes the entire spectrum of the bulb when projecting grays instead of projecting the superposition of light from the red, green, and blue filters. This makes these DLP projectors behave like four-primary devices (like printers that use cyan, magenta, yellow, and black as primaries). Adding the fourth primary results in linear dependency of the fourth primary on the other three. Thus, a color cannot be represented using unique combinations of the four primaries. In fact, the shape of the gamut depends not only on the primaries used but also on the algorithms used to produce different colors by combining these primaries.

The gamut-matching method described above uses a parametric definition of the color gamuts as vector addition of three primaries and thus can only be applied to systems with three linearly independent primaries. Wallace et al. [101] present a nonparametric version of the same method that can be applied to DLP projectors with a clear filter. In four-primary systems, the 3D color gamut is no longer a rectangular parallelepiped. Based on the different kinds of color-combination algorithms used, different 3D color gamuts can result. The nonparametric gamut-matching algorithm follows the same three steps of reconstructing the 3D gamut and identifying and achieving a common 3D gamut, but it uses a nonparametric representation of the 3D color gamut.

1. Reconstruction. Since the 3D gamut cannot be parameterized, one should ideally measure the XYZ-coordinates of all of the 2^{24} RGB inputs to find the 3D gamut. However, this is not feasible. So, the

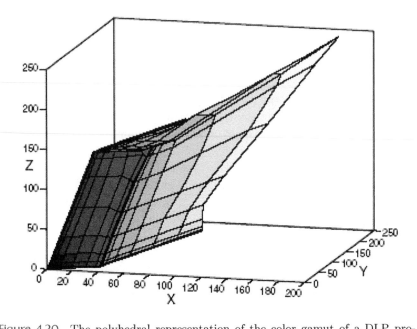

Figure 4.20. The polyhedral representation of the color gamut of a DLP projector that uses four primaries. (From [101]. © Eurographics Association 2003. Reproduced by kind permission of the Eurographics Association.)

RGB input volume is sampled uniformly, and the XYZ-coordinates of the sampled inputs are measured using a radiometer. A triangle mesh is then generated from the sampled data to form a polyhedral representation of the 3D gamut. Such a reconstructed gamut for a DLP projector is shown in Figure 4.20. Note that this gamut is no longer convex in shape as gamuts of three-primary systems are. Instead, it shows marked concavities.

2. Modification. A geometric algorithm is used to find the intersection of the gamuts of all of the projectors; however, since the gamuts can be concave, the intersection operation might produce a set of disjoint polyhedra. In this case, the polyhedron with the largest volume is used as the common 3D gamut to which every projector should transform.

3. Reprojection. The final step is to find a 3D color mapping for each projector that modifies every RGB input to a new RGB input that samples the common gamut uniformly. This is achieved in two steps.

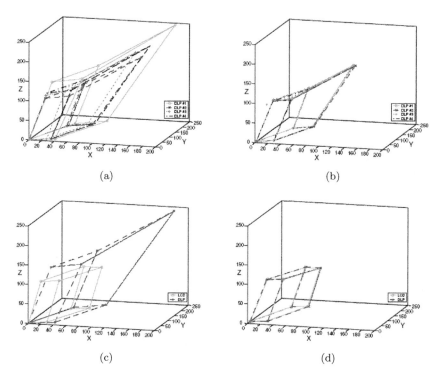

(a)   (b)

(c)   (d)

Figure 4.21. The gamuts of four DLP projectors (a) before and (b) after nonparametric gamut matching. The gamuts of a DLP and LCD projector (c) before and (d) after parametric gamut matching. Note that since the LCD projector is a three-primary system, its gamut is a regular convex parallelepiped. (From [101]. © Eurographics Association 2003. Reproduced by kind permission of the Eurographics Association.)

First, a standard color-transfer function $M(r, g, b)$ is designed that defines the colors for display in the common gamut C as the input (r, g, b) changes. Let the initial functions that defined the projected color for each projector i in the original gamut be defined by $M_i(r_i, g_i, b_i)$. Then the common color gamut is achieved by a function $M_i^{-1}M$. Note that unlike a three-primary system where M and M_i are both linear parametric functions represented by matrices, in this case M and M_i are both nonlinear and nonparametric. Hence, $M_i^{-1}M$ is implemented by a per projector 3D color map. This method, although designed for DLP projectors, can also be used to calibrate a mix of different kinds of projectors, even if some of them use a three-primary system. Figure 4.21 shows some results [101].

Gamut matching ignores the intra-projector spatial variation. As a result, spatially varying parameters such as Q_l and c_l are not sampled adequately in the spatial domain. So, gamut-matching methods cannot address the spatial color variations. Furthermore, expensive instrumentation makes this method cost-prohibitive. A relatively inexpensive radiometer costs at least four times as much as a projector. Expensive radiometers can cost as much as a dozen projectors. Finally, the algorithm to find a common color gamut for n projectors, even for linear systems, is a computational-geometry problem of complexity $O(n^6)$ [4], which makes this method not scalable for displays made up of large numbers of projectors.

To avoid this, another method [60] makes the assumption that chrominance across a display made up of the *same model* of projectors does not vary significantly. Then, the problem of gamut matching reduces to matching the per-channel ITF of all of the projectors to a linear function within a range of intensity that can be produced by all projectors. Such a range is simple to find. The maximum intensity is given by the minimum of the maximum channel intensity of all of the projectors, and the minimum intensity is given by the maximum of the minimum channel intensity of all projectors. Thus, under the assumption of no intra-projector spatial variation, this method achieves *strict photometric uniformity*.

4.4.3 Blending

Blending or feathering techniques skip the reconstruction and the modification steps and directly apply a reprojection step, i.e., a modification of inputs [79, 80, 103]. These modifications, adopted from image-mosaicing techniques, address overlapped regions and try to smooth color transitions across these regions. The smooth transitions can be achieved by using a linear or cosine ramp, which attenuates pixel intensities in the overlapped region. For example, consider a pixel x in the overlap region of two projectors P_1 and P_2. Let the contributions of these projectors at x be given by $P_1(x)$ and $P_2(x)$, respectively. When using linear ramping, the intensity at x is computed by a linear combination of the intensities $P_1(x)$ and $P_2(x)$, i.e,

$$\alpha_1(x)P_1(x) + \alpha_2(x)P_2(x),$$

where $\alpha_1 + \alpha_2 = 1$. The weights α_1 and α_2 are chosen based on the distance of x from the boundaries of the overlapped region. During geometric calibration, the image of the projectors is captured by a camera. These distances are computed in the camera's coordinate space. For example, when using a linear ramp, if the distances of x from the boundary of P_1

and P_2 are d_1 and d_2, respectively, α_1 and α_2 are chosen as follows:

$$\alpha_1(x) = \frac{d_1}{d_1 + d_2}; \quad \alpha_2(x) = \frac{d_2}{d_1 + d_2}.$$

This two-projector example can be extended to an arbitrary number of projectors [80]. To do so, first the convex hull H_i of all of the pixels in projector P_i is computed in the camera coordinate space. This defines an area in the camera coordinate space where pixels from P_i are present. The alpha weight $A_m(x)$ associated with projector P_m's pixel x is then evaluated as follows:

$$A_m(x) = \frac{\alpha_m(m, x)}{\sum_i \alpha_i(m, x)}, \tag{4.13}$$

where $\alpha_i(m, x) = w_i(m, x) \cdot d_i(m, x)$ and i is the index of the projectors observed by the camera (including projector m). In this equation, $w_i(m, x) = 1$ if x from P_m is inside H_i in the camera coordinate space. This indicates that although x is from a projector P_m that is different from projector P_i, it falls in the overlap region of P_i and P_m. If $w_i(m, x) = 1$, then $d_i(m, x)$ is the distance from x to the nearest edge of H_i in the camera coordinate space. If x is not within H_i, then it does not overlap with P_i and there is no need to consider P_i when calculating A_m. So, in this case, $w_i(m, x) = 0$. The alpha masks thus generated are applied after the image has been warped to achieve geometric calibration.

Figure 4.22 shows the alpha masks created for four overlapping projectors. Note that the alpha masks produce edge artifacts. These artifacts occur along the diagonals of each projector's coordinate system, where the choice of the nearest edge of H_i changes. Essentially, discontinuities in the first-order spatial derivative of d_m result here and propagate to form similar discontinuities in A_m. The discontinuity inside one projector's alpha mask is complemented by similar ones within all masks of overlapping projectors, and these must be very precisely aligned by geometric correction to avoid creating edge artifacts, to which the human visual system is highly attuned. A recent work presents a new method for creating these alpha masks that avoids such discontinuities [36]. This is achieved effectively by just changing the summation in the denominator of Equation (4.13) to a product. Thus, the new alpha mask equation is

$$A_m(x) = \frac{\alpha_m(m, x)}{\prod_i \alpha_i(m, x)}.$$

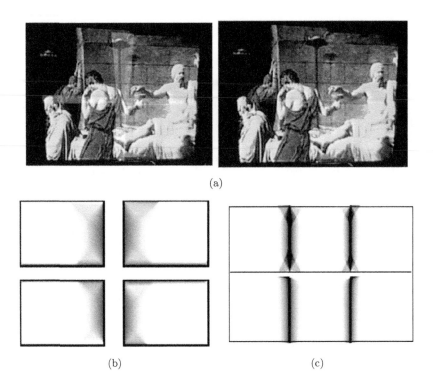

Figure 4.22. (a) Results of software blending on a four-projector display. (b) The alpha masks computed for four projectors for this purpose. (c) This compares the regular alpha masking with the smoother version where the summation in the denominator is replaced with a product. The top image shows the alpha masks of three projectors in a 3×1 array of projectors. The bottom image shows the smoother alpha masks for the same multi-projector system. (© 2002 IEEE. Reprinted, with permission, from [12].)

It can be shown that the isocontours of this function form smooth hyperbolae within the projector coordinate system, approaching zero at the projector boundaries. This function is thus suitable for generating smoother alpha masks, as shown in Figure 4.22.

Until now, we have discussed achieving blending in software. In this case, the ramps in the overlap region can be precisely controlled by software. However, this cannot attenuate the black offset, especially important with scientific or astronomical data, which often have black backgrounds. Alternate optical methods try to achieve this blending by physical attenuation of lights so that it can also affect the black offset. In one method, physical masks mounted at the projector boundaries on the optical path

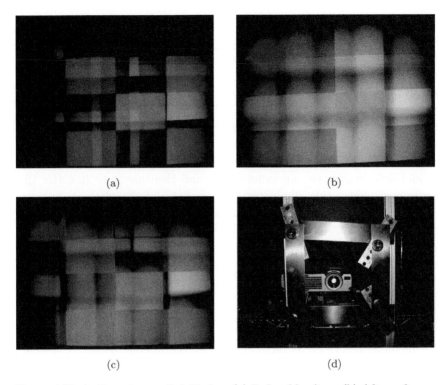

Figure 4.23. A 15-projector tiled display. (a) Before blending. (b) After software blending; note the French-door effect because the projectors were not linearized. (c) After aperture blending. (d) The hardware used for aperture blending where metal masks are mounted on the optical path of the projector that attenuates the light physically. (From [58]. © 2005 ACM, Inc. Included here by permission. Display courtesy of Argonne National Laboratory.)

attenuate the light in the overlapped region [52], as shown in Figure 4.23. In another method, optical masks are inserted in front of the projection lens to achieve the attenuation [18].

Although blending methods are automated and scalable, they do not reconstruct the function E, even in the overlap region. Thus, they ignore the inter- and intra-projector spatial color variation, and the variation in the overlapped region is not accurately estimated. Each overlap region defines a zone of transition from one projector to another, and this method aims at making this transition perceptually smooth. Thus, the formal goal is to generate *perceptual color uniformity*. However, the size of the overlap region over which the intensity changes from one value to another is fixed. To achieve a smoothness that is imperceptible to the human eye, the size needs to be controlled based on the difference in the intensity of the projectors sharing the overlap region [17]. So, blending works well only if the overlapping projectors have similar intensity ranges. This is often assured by an initial manual intensity adjustment using the projector controls. However, for displays where the intensity has a large spatial variation, since E is not estimated accurately, blending results in only softening of the seams in the overlapping region, rather than removing them completely. Finally, the multiplication of inputs from projectors by the alpha masks inherently assumes a linear input transfer function. So, a correct blending algorithm should first recover the projector's ITF and then use it to linearize the projectors. Since this eliminates the simplicity of the method, in most scenarios this linearization step is not performed. In some cases, if the projector controls allow control of the gamma function, the gamma is manually preset to be linear. If this linearization is not performed, blending shows dark bands in the overlap region due to overcompensation. This creates what we call a "French-door effect," as illustrated in Figure 4.23.

4.4.4   Camera-Based Photometric Seamlessness

All methods described so far address only the inter-projector or overlap variation. None addresses the intra-projector variation, which can be significant. Also, only the gamut-matching method makes an effort to estimate the entire color response of the projectors at a limited resolution. Since the spatial variation in color is significant, a high-resolution estimation of the color response is the only means towards an accurate solution. So, the use of a camera is inevitable, but since a camera has a limited color gamut (as opposed to a spectroradiometer), estimating the color gamut of the display at a high resolution is difficult. However, different exposure settings of a camera can be used to measure the photometric (intensity) response of the multi-projector display accurately and faithfully. Exploiting this fact, one can use a camera to correct for the photometric variation across a multi-projector display [56, 57, 58]. Since most current displays use the same model of projectors that have similar chrominance properties, this method achieves reasonable seamlessness. The method follows the color-variation model in Section 4.3, so we suggest that the reader review that section briefly before advancing further.

This technique makes some assumptions to simplify the color-variation model considerably. It assumes that the projection surface is Lambertian in nature and hence does not have any view dependency. From Equation (4.6), the intensity projected at any display coordinate (u, v) for an input $i =$

(i_r, i_g, i_b) is given by

$$E(u, v, i, e) = T(s, t, i).$$

This does not depend on Λ anymore and only addresses intensity. By removing the chrominance component of Equation (4.4), the intensity projected by each channel l is given by

$$\text{int}(D_l(s, t, i_l)) = h(i_l)Q_l(s, t).$$

So, the intensity $L(s, t, i)$ projected by the three-channel input (i_r, i_g, i_b) is derived from Equation (4.5) as

$$L(s, t, i) = \text{int}(T(s, t, i)) = D_r(s, t, i) + D_g(s, t, i) + D_b(s, t, i) + B(s, t, i).$$
$$(4.14)$$

In the color-variation model described in Section 4.3, $Q_l(s, t)$ denotes the spatial intensity profile of channel l considering the perfect projection system where there is no black offset. Since it is not possible to remove the black offset physically from the projection, it is not possible to measure $Q_l(s, t)$ directly. We define $W_l(s, t)$ as the maximum intensity projected by channel l at (s, t). Now, $Q_l(s, t)$ can be defined as

$$Q_l(s, t) = W_l(s, t) - B(s, t). \tag{4.15}$$

We call W_l the *channel-intensity profile* and B the *black-intensity profile*. Replacing Q_l in Equation (4.14) with Equation (4.15), we get

$$L(s, t, i) = \left(\sum_{l \in \{r, g, b\}} h_l(i_l)(W_l(s, t) - B(s, t)) \right) + B(s, t). \tag{4.16}$$

The black- and channel-intensity profiles are together called *projector-intensity profiles*.

Under these conditions, the equation for the multi-projector display, Equation (4.7) becomes

$$L(u, v, i) = \text{int}(E(u, v, i)) = \sum_{j \in N_P} L(s_j, t_j, i), \tag{4.17}$$

where $(u, v) = G_j(s_j, t_j)$ by Equation (4.3) and N_P is the set of projectors overlapping at display coordinate (u, v). The variation in L with respect to (u, v) in Equation (4.17) describes the *photometric variation* in a multi-projector display.

Reconstruction. There are *three* projector parameters on which L in Equation (4.17) depends: the transfer functions h_l and the intensity profiles W_l and B. In this step, we describe the reconstruction of these parameters for each projector.

- Transfer function (h_l). Since $h_l(i_l)$ is spatially constant [62], it can be estimated at low resolution, even at only one spatial coordinate for each projector, supporting the use for a photometer for this measurement. Although measuring using a photometer is a slow process (1–20 seconds per measurement), it can be justified since h_l changes little temporally [62] and needs to be measured very infrequently, usually once in 9–12 months.

  However, photometers are expensive, so it would be ideal if h_l were estimated by a camera. This can be achieved using high dynamic range imaging methods [25, 78], where each intensity level projected from the projector is captured at different exposures. These images are then processed to find the relative intensity corresponding to each input in a high dynamic range scale from which the transfer function h_l can be easily estimated. Since h_l is spatially invariant, only a small region at the center of the projector is analyzed. Since this small region is in the non-overlapping region of each projector, h_l for all projectors can be estimated in parallel. Essentially, all of the projectors can be turned on together during data capture.

  The process to estimate h_g is described first. Other transfer functions, h_r and h_b, are estimated analogously. Input $i_G = (0, i_g, 0)$, $0.0 \leq i_g \leq 1.0$ is defined as one where the green channel input is i_g and the red and blue channel inputs are 0. The intensity response $L(s, t, i_G)$ for $O(2^8)$ values of i_g is measured, and the input space of channel g is densely sampled. Since this intensity response is measured at one spatial location, we omit the spatial coordinates and denote it as $L(i_G)$ only.

  The measured intensity $L(i_G)$ includes the black offset and is not monotonic with respect to i_G. So, the *maximum input range* $i_{G_m} \leq i_G \leq i_{G_M}$ within which $L(i_G)$ is monotonic is found. The inputs $i_{G_m} = (0, i_{g_m}, 0)$ and $i_{G_M} = (0, i_{g_M}, 0)$ denote, respectively, the lower and the upper boundary of this range. The transfer function h_g is then estimated from Equation (4.16) as

$$h_g(i_g) = \frac{L(i_G) - L(i_{G_m})}{L(i_{G_M}) - L(i_{G_m})}.$$

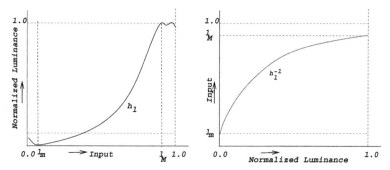

Figure 4.24. The transfer function (left) and the inverse transfer function (right) for a single channel of a projector. (From [58]. © 2005 ACM, Inc. Included here by permission.)

The function h_g estimated in this matter is monotonic with $\min_{\forall i_g} h_g(i_g) = h_g(i_{g_m}) = 0$ and $\max_{\forall i_g} h_g(i_g) = h_g(i_{g_M}) = 1$ (see Figure 4.24).

- Intensity profiles (W_l, B). W_l and B are spatially varying functions. A digital camera can be used to estimate them [56, 57]. First, for a suitable position and orientation of the camera looking at the entire display, the geometric-warping function that transforms every projector pixel to the appropriate camera pixel is estimated using any camera-based geometric-calibration method as described in Chapter 3. After geometric registration, an appropriate test image is projected by the projector and captured by the camera, and the intensity profiles are estimated. The images from more than one non-overlapping projector can be captured in the same image. Note that the images to estimate these functions must be taken from the same position and orientation as the images for geometric calibration.

The test images are comprised of identical input at every projector coordinate. Equation (4.16) shows that the input to be combined to estimate B is $(i_{r_m}, i_{g_m}, i_{b_m})$, that is, the input that projects minimum intensity for each channel. Similarly, to estimate the maximum-intensity function for the green channel, W_g, $(i_{r_m}, i_{g_M}, i_{b_m})$ should be projected, that is, the input that projects maximum intensity from the green channel and minimum intensity from the other two channels. The test images for estimating W_g of all the projectors for a four- and 15-projector display are shown in Figure 4.25. Figure 4.17(b) shows W_g and B thus estimated for one of these projectors. The

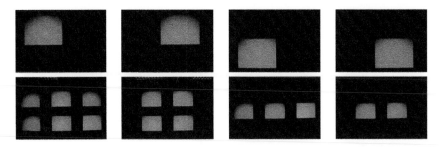

Figure 4.25. To compute the maximum intensity function for the green channel of each projector, we need only four pictures. This reduction in the number of images is achieved by turning on more than one non-overlapping projector while capturing each image. Top: Pictures taken for a display made up of a 2×2 array of four projectors. Bottom: The pictures taken for a display made up of a 3×5 array of 15 projectors. (From [58]. © ACM, Inc. Included here by permission.)

computation of test images for the estimation of W_r and W_b are done analogously.

Modification. The first goal that comes to mind for achieving a photometrically seamless display is that of strict photometric uniformity. Essentially, the desired function should produce a uniform intensity response at every pixel for any input.

For this, parameters that are analogous to the parameters h_l, W_l, and B for a single-projector display are defined for a multi-projector display. Unlike a single-projector display where h_l is spatially constant, different parts of the multi-projector display have different h_l since they are projected from different projectors. As a result, functions analogous to the intensity profiles (W_l and B) of a single projector cannot be defined for a multi-projector display. Therefore, a perceptually appropriate *common transfer function* $\mathcal{H}_l$ is chosen that is spatially invariant throughout the multi-projector display. This enables the *identification of display intensity profiles* $\mathcal{W}_l$ and $\mathcal{B}$ for the entire multi-projector display. Finally, *flattening of the display intensity profiles* is performed to remove all intensity variation and achieve a uniform intensity response across the display and, hence, photometric uniformity. Following are the detailed descriptions of the steps involved.

1. Choosing a common transfer function. A *common transfer function* $\mathcal{H}_l$ for each channel l should satisfy the following three conditions: $\mathcal{H}_l(0) = 0$, $\mathcal{H}_l(1) = 1$, and $\mathcal{H}_l(i_l)$ is monotonic. The transfer

functions h_l of all the different projectors are then replaced by this common $\mathcal{H}_l$ to assure a spatially invariant transfer function across the whole multi-projector display. So, Equation (4.8) becomes

$$L(u, v, i) = \sum_l \left(\mathcal{H}_l(i_l) \sum_j \left(W_{l_j}(s_j, t_j) - B_j(s_j, t_j) \right) \right) + \sum_j B_j(s_j, t_j),$$
$$(4.18)$$

where W_{l_j} and B_j are the intensity profiles for projector P_j.

We use $\mathcal{H}_l = i_l^2$, which is commonly used to approximate the logarithmic response of the human eye to varying intensity. If $\mathcal{H}_l$ is chosen to be a linear function, it results in washed-out images.

2. Identifying the display-intensity profiles. In Equation (4.18), using

$$\mathcal{W}_l(u, v) = \sum W_{l_j}(s_j, t_j)$$

and

$$\mathcal{B}(u, v) = \sum B_j(s_j, t_j),$$

with appropriate coordinate transformation between projector and display coordinate space, we get

$$L(u, v, i) = \left(\sum_{l \in \{r, g, b\}} \mathcal{H}_l(i_l) \left(\mathcal{W}_l(u, v) - \mathcal{B}(u, v) \right) \right) + \mathcal{B}(u, v). \quad (4.19)$$

Note that the above equation for the *whole multi-projector display* is exactly analogous to that for a single projector (Equation (4.16)). Thus, we call $\mathcal{W}_l$ and $\mathcal{B}$ the *channel display-intensity profile* for channel l and the *black display-intensity profile*, respectively. Figure 4.26 shows $\mathcal{W}_g$ for a four-projector display and $\mathcal{W}_g$ and $\mathcal{B}$ for a 15-projector display.

3. Modifying the display-intensity profiles. The multi-projector display defined by Equation (4.19) still does not look like a single-projector display because, unlike the single-projector intensity functions (Figure 4.17), the analogous multi-projector intensity functions (Figure 4.26) have sharp discontinuities that result in perceivable seams in the display. Since the transfer function is already identical

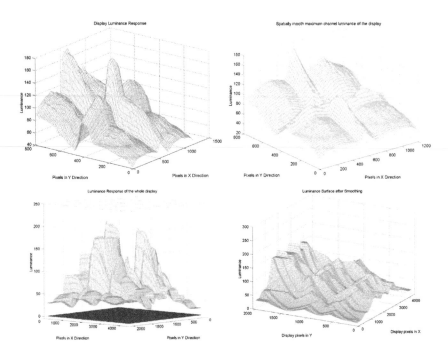

Figure 4.26. The estimated maximum display-intensity function of the green channel ($\mathcal{W}_g$) for a 2 × 2 array of projectors (top left). The estimated maximum display-intensity function for the green channel ($\mathcal{W}_g$) and the black display-intensity function ($\mathcal{B}$) of a 3 × 5 array of projectors (bottom left). The high-intensity regions in both correspond to the overlap regions across different projectors. The graphs in the top right and bottom right show the smooth maximum display intensity function for the green channel ($\mathcal{W}_g'$) achieved by applying the constrained gradient-based smoothing algorithm on the maximum display-intensity functions in the top left and bottom left, respectively. Note that these are perceptually smooth even though they are not geometrically smooth. (From [58]. © 2005 ACM, Inc. Included here by permission.)

at every pixel of the display, the spatial variation in the intensity profiles are the sole cause of the photometric variation at this point. So, to achieve strict photometric uniformity, $\mathcal{W}_l(u, v)$ and $\mathcal{B}(u, v)$ are flattened to generate $\mathcal{W}_l'(u, v)$ and $\mathcal{B}'(u, v)$, respectively. These are called the *flat display-intensity functions*. The functions $\mathcal{W}_l'(u, v)$ and $\mathcal{B}'(u, v)$ are chosen to match the most restricted pixels, i.e., the pixels with the dimmest maximum and the brightest minimum respectively. This assures that the flat profiles are achievable by all of the pixels

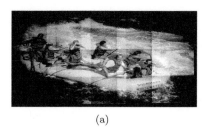
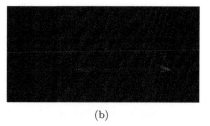

(a)   (b)

Figure 4.27. Digital photographs of a 15-projector tiled display (8 feet × 10 feet in size) (a) before any correction and (b) after strict photometric uniformity where the display has a very low dynamic range. (From [58]. © 2005 ACM, Inc. Included here by permission. Display courtesy of Argonne National Laboratory.)

of the display. Mathematically,

$$\mathcal{W}_l'(u, v) = \min_{\forall(u,v)} \mathcal{W}_l(u, v),$$

$$\mathcal{B}'(u, v) = \max_{\forall(u,v)} \mathcal{B}(u, v).$$

The photometrically uniform function $L'(u, v, i)$ thus generated, derived from Equation (4.17) is

$$L'(u, v, i) = \left(\sum_{l \in \{r,g,b\}} \mathcal{H}_l(i_l) \left(\mathcal{W}_l'(u, v) - \mathcal{B}'(u, v) \right) \right) + \mathcal{B}'(u, v).$$

When each of $\mathcal{W}_l'(u, v)$ and $\mathcal{B}'(u, v)$ is flat and the transfer function $\mathcal{H}_l$ is spatially invariant, $L(u, v, i) = \text{int}(E'(u, v, i))$ satisfies the goal of photometric uniformity (4.9). However, since the intensity at every pixel is matched to the pixel with the most limited dynamic range, a shortcoming of this method is severe compression in the dynamic range of the display, as illustrated in Figure 4.27. This compression in the dynamic range is especially pronounced for displays with inexpensive commodity projectors where it can render them almost useless.

This compression of the dynamic range makes one question the goal of strict photometric uniformity. In fact, if we go back and analyze the properties of a practical single projector once more, we find that although the transfer function is indeed spatially invariant [62], the intensity profiles of W_l and B are hardly flat, as shown in Figure 4.17, but humans perceive the image projected by a single projector as uniform. Thus, it is evident from the very case of single projectors that strict photometric uniformity is not required for a seamless display; rather, single projectors have a smoothly varying response that is imperceptible to the human eye ensuring a *perceptual* photometric uniformity, as defined in Equation (4.10). Therefore, it is probably sufficient to design a multi-projector display that is made to look like a single-projector display, such that one cannot discern the number of projectors making up the display.

Note that the criterion of perceptual uniformity alone may not ensure a good display quality, i.e., high intensity and contrast. For example, strict photometric uniformity, which implies perceptual uniformity, does not ensure good display quality since it forces the display quality to match the "worst" pixel on the display, leading to compression in the dynamic range. However, the goal of perceptual uniformity, being less restrictive, can provide the extra leverage to increase the overall display quality. The problem of achieving photometric seamlessness can be defined as an optimization problem where the goal is to achieve perceptual uniformity while maximizing the display quality. This optimization is a general concept and can have different formalizations based on two factors: the parameters on which δ of Equation (4.10) depends, and the way display quality is defined.

Perception studies show that humans are sensitive to significant intensity discontinuities but can tolerate smooth intensity variations [22, 34, 99]. To make the display-intensity profiles perceptually uniform, $\mathcal{W}_l(u,v)$ and $\mathcal{B}(u,v)$ are made smooth (not flat) to generate $\mathcal{W}_l'(u,v)$ and $\mathcal{B}'(u,v)$, respectively. These are called the *smooth display-intensity profiles.*

$\mathcal{B}'(u,v)$ is approximated as

$$\mathcal{B}'(u,v) = \max_{\forall u,v} \mathcal{B}(u,v),$$

since the variation in $\mathcal{B}(u,v)$ is almost negligible when compared to $\mathcal{W}_l(u,v)$ (Figure 4.26). The function $\mathcal{B}'(u,v)$ thus defines the minimum intensity that can be achieved at all display coordinates.

The smoothing of $\mathcal{W}_l(u,v)$ is formulated as an *optimization problem* where a $\mathcal{W}_l'(u,v)$ that minimizes the deviation from the original $\mathcal{W}_l(u,v)$ assuring high dynamic range and at the same time maximizes its smoothness assuring perceptual uniformity is devised. The smoothing criteria are chosen based on quantitative measurement of the limitation of the human eye in perceiving smooth intensity variations. A *constrained gradient-based smoothing* method finds the optimal solution to this problem. To generate $\mathcal{W}'_l(u,v)$ from $\mathcal{W}_l(u,v)$, this technique uses the following optimization constraints.

- **Capability constraint.** This constraint of $\mathcal{W}'_l \leq \mathcal{W}_l$ ensures that $\mathcal{W}'_l$ never goes beyond the maximum intensity $\mathcal{W}_l$ achievable by the display. In practice, with discrete sampling of these functions,

$$\mathcal{W}'[u][v] < \mathcal{W}[u][v], \qquad\qquad \forall u, v.$$

- **Perceptual-uniformity constraint.** This constraint assures that $\mathcal{W}'_l$ has a smooth variation imperceptible to humans:

$$\frac{\partial \mathcal{W}'_l}{\partial x} \leq \frac{1}{\lambda} \times \mathcal{W}'_l,$$

where λ is the *smoothing parameter* and $\frac{\partial \mathcal{W}'_l}{\partial x}$ is the gradient of $\mathcal{W}'_l$ along any direction x. Compare this inequality with Inequality (4.10). In the discrete domain, when the gradient is expressed as a linear filter involving the eight neighbors (u', v') of a pixel (u, v), $u' \in \{u - 1, u, u + 1\}$ and $v' \in \{v - 1, v, v + 1\}$, this constraint is given by

$$\frac{|\mathcal{W}'[u][v] - \mathcal{W}'[u'][v']|}{\sqrt{|u - u'|^2 + |v - v'|^2}} \leq \frac{1}{\lambda}\mathcal{W}'[u][v], \ \forall u, v, u', v'.$$

- **Display-quality objective function.** The above two constraints can yield many feasible $\mathcal{W}'_l$. To maximize dynamic range, the integration of $\mathcal{W}'_l$ has to be maximized. In the discrete domain, this is expressed as

$$\text{maximize } \sum_{u=0}^{X-1} \sum_{v=0}^{Y-1} \mathcal{W}'[u][v],$$

where X and Y denote the height and width of the multiprojector display in number of pixels.

A dynamic-programming method provides the optimal solution for this optimization in linear time with respect to the number of pixels in the display, i.e., $O(XY)$. The time taken to compute this solution on an Intel Pentium III 2.4 GHz processor for displays with 9 million pixels is less than one second. The pseudocode for the algorithm is as follows:

$\forall (u, v),$ $\qquad\qquad\qquad \mathcal{W}'(u, v) \leftarrow \mathcal{W}(u, v);$
$\delta \leftarrow \frac{1}{\lambda};$

for $u = 0$ **to** $X - 1$
   for $v = 0$ **to** $Y - 1$
      $\mathcal{W}'(u, v) \leftarrow \min(\mathcal{W}'(u, v), (1 + \sqrt{2}\delta)\mathcal{W}'(u - 1, v - 1),$
             $(1 + \delta)\mathcal{W}'(u - 1, v), (1 + \delta)\mathcal{W}'(u, v - 1));$

for $u = X - 1$ **down to** 0
   for $v = 0$ **to** $Y - 1$
      $\mathcal{W}'(u, v) \leftarrow \min(\mathcal{W}'(u, v), (1 + \sqrt{2}\delta)\mathcal{W}'(u + 1, v - 1),$
             $(1 + \delta)\mathcal{W}'(u + 1, v), (1 + \delta)\mathcal{W}'(u, v - 1));$

for $u = 0$ **to** $X - 1$
   for $v = Y - 1$ **down to** 0
      $\mathcal{W}'(u, v) \leftarrow \min(\mathcal{W}'(u, v), (1 + \sqrt{2}\delta)\mathcal{W}'(u - 1, v + 1),$
             $(1 + \delta)\mathcal{W}'(u - 1, v), (1 + \delta)\mathcal{W}'(u, v + 1));$

for $u = X - 1$ **down to** 0
   for $v = Y - 1$ **to** 0
      $\mathcal{W}'(u, v) \leftarrow \min(\mathcal{W}'(u, v), (1 + \sqrt{2}\delta)\mathcal{W}'(u + 1, v + 1),$
             $(1 + \delta)\mathcal{W}'(u + 1, v), (1 + \delta)\mathcal{W}'(u, v + 1));$

Note that the smoothing achieved likewise is significantly different from the linear smoothing function usually used in the domain of image processing [35, 50, 51]. For example, a gradient- or curvature-based linear smoothing filter, which are popular operators in image-processing applications, smooths the hills and fills the troughs. However, the constraints of the smoothing of the channel-intensity profile are such that while the hills can be smoothed, the troughs cannot be filled since the response thus achieved will be beyond the display capability of the projectors.

Figure 4.26 shows $\mathcal{W}'_g(u, v)$ for a 15-projector display. The perceptually seamless intensity response of the display $L'(u, v, i)$ thus generated, derived from Equation (4.19), is

$$L'(u, v, i) = \left(\sum_{l \in \{r, g, b\}} \mathcal{H}_l(i_l) \left(\mathcal{W}'_l(u, v) - \mathcal{B}'(u, v) \right) \right) + \mathcal{B}'(u, v).$$

When each of $\mathcal{W}'_l(u,v)$ and $\mathcal{B}'(u,v)$ is smooth and the transfer function $\mathcal{H}_l$ is spatially invariant, $L'(u,v,i)$ is also smooth and satisfies Inequality (4.10).

The results of this method to display walls of different sizes are illustrated in Figures 4.28, 4.29, and 4.30. They show the digital photographs of results on a 2×3 array of six projectors (4 feet wide and 3 feet high), a 3×5 array of 15 projectors (10 feet wide and 8 feet high), and a 2×2 array of four projectors (2.5 feet wide and 1.5 feet high).

The smoothing parameter λ is derived from the human contrast sensitivity function (CSF) [99, 22]. Contrast threshold defines the minimum percentage change in intensity that can be detected by a human being at a particular spatial frequency. The human CSF is a plot of the contrast sensitivity (reciprocal of contrast threshold) with respect to spatial frequency (Figure 4.31). The CSF is bow-shaped with maximum sensitivity at a frequency of 5 cycles per degree of angle subtended on the eye. In other words, a variation of less than 1% in intensity will be imperceptible to humans for a spatial grating of 5 cycles/degree. For other frequencies, a greater variation can be tolerated. This fact is used to derive the relative intensity that a human being can tolerate for every pixel of the display as follows.

Let d be the perpendicular distance of the user from the display (can be estimated by the position of the camera used for calibration), r the resolution of the display in pixels per unit distance, and τ the contrast threshold that humans can tolerate per degree of visual angle (1% at peak sensitivity). The number of pixels subtended per degree of the human eye is given by $\frac{d\pi r}{180}$. Since the peak sensitivity occurs at 5 cycles per degree of visual angle, the number of display pixels per cycle of the grating is then given by $\frac{d\pi r}{180 \times 5}$. Within this pixel range, an intensity variation of τ will go undetected by humans. Thus,

$$\lambda = \frac{d\pi r}{900\tau}. \tag{4.20}$$

For a 15-projector display with r of about 30 pixels per inch, for a user about 6 feet away from the display, by substituting $\tau = 0.01$ in Equation (4.20), we get a λ of about 800. Note that as the user moves farther away from the display, λ increases up, i.e., the surface needs to be smoother and will have lower dynamic range. This

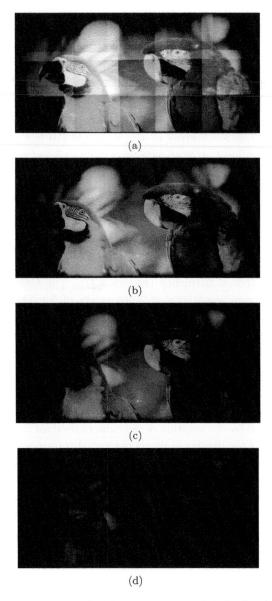

(a)

(b)

(c)

(d)

Figure 4.28. Digital photographs of a 15-projector tiled display (8 feet × 10 feet in size) (a) before any correction and after constraint gradient-based smoothing with smoothing parameter of (b) $\lambda = 400$, (c) $\lambda = 800$, and (d) $\lambda = \infty$. Note that the dynamic range of the display reduces as the smoothing parameter increases. At $\lambda = \infty$, we have the special case of photometric uniformity where the display has a very low dynamic range. (From [58]. © 2005 ACM, Inc. Included here by permission. Display courtesy of Argonne National Laboratory.)

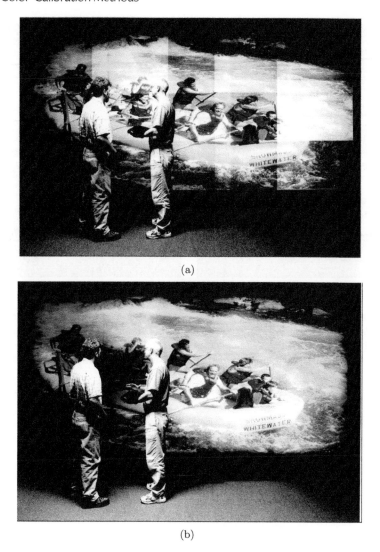

(a)

(b)

Figure 4.29. Digital photographs of a display made up of a 3 × 5 array of 15 projectors (10 feet wide and 8 feet high). (a) Before correction. (b) Perceptual photometric seamlessness after applying constrained gradient-based intensity smoothing. (From [58]. © 2005 ACM, Inc. Included here by permission. Display courtesy of Argonne National Laboratory.)

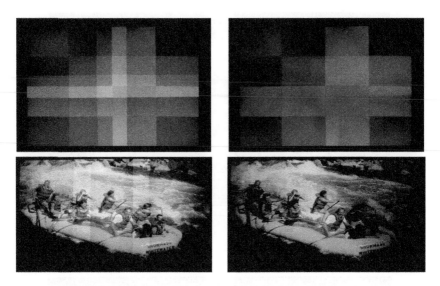

Figure 4.30. Digital photographs of displays made up of a 2 × 2 and a 2 × 3 array of four (1.5 feet × 2.5 feet in size) and six (3 feet × 4 feet in size) projectors, respectively, before correction (left) and after constrained gradient-based intensity smoothing (right). (From [58]. © 2005 ACM, Inc. Included here by permission. Display courtesy of Argonne National Laboratory)

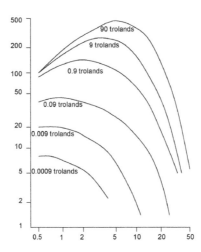

Figure 4.31. The contrast sensitivity function (CSF) of the human eye for decreasing luminance changes from a band-pass filter to a low-pass filter.

also explains the variations perceptible in some of the images in Figures 4.28, 4.29, 4.30, and 4.34 even though they are imperceptible when seen in-person on the large display.

Reprojection. In the previous section, the display parameters were theoretically modified to generate a hypothetical display that is photometrically uniform. The practical display needs to behave like the hypothetical display with these desired parameters. The projector hardware does not offer the precision control to modify $\mathcal{W}_l(u,v)$, $\mathcal{B}(u,v)$, and h_l directly to create the hypothetical display. Consequently, these modifications are achieved by modifying the input i_l at every projector coordinate. In this section, we explain how this modification is achieved for *any one projector* of a multi-projector display at its coordinate (s,t). Since the modification is pixel-based, we have left out the (s,t) from all of the equations in this ssection.

For a given i_l, the actual response of the display is given by

$$\sum_{l\in\{r,g,b\}} (h_l(i_l)(\mathcal{W}_l - \mathcal{B})) + \mathcal{B}.$$

The goal of reprojection is to get the response

$$\sum_{l\in\{r,g,b\}} (\mathcal{H}_l(i_l)(\mathcal{W}_l' - \mathcal{B}')) + \mathcal{B}'$$

to simulate a modified perceptually seamless display. So, the input i_l is modified to i_l' such that

$$\sum_{l\in\{r,g,b\}} (\mathcal{H}_l(i_l)(\mathcal{W}_l' - \mathcal{B}')) + \mathcal{B}' = \sum_{l\in\{r,g,b\}} (h_l(i_l')(\mathcal{W}_l - \mathcal{B})) + \mathcal{B}. \tag{4.21}$$

It can be shown that the following i_l' solves Equation (4.21).

$$i_l' = h_l^{-1}(\mathcal{H}_l(i_l)\mathcal{S}_l + \mathcal{O}_l), \tag{4.22}$$

where h_l^{-1} is the inverse transfer function of a projector and can be computed directly from h_l as shown in Figure 4.24, and $\mathcal{S}_l(u,v)$ and $\mathcal{O}_l(u,v)$ are called the *display scaling map* and *display offset map*, respectively, and are given by

$$\mathcal{S}_l(u,v) = \frac{\mathcal{W}_l'(u,v) - \mathcal{B}'(u,v)}{\mathcal{W}_l(u,v) - \mathcal{B}(u,v)}; \ \mathcal{O}_l(u,v) = \frac{\mathcal{B}'(u,v) - \mathcal{B}(u,v)}{3(\mathcal{W}_l(u,v) - \mathcal{B}(u,v))}.$$

(a)   (b)

Figure 4.32. (a) The scaling map for a display made up of a 5×5 array of 15 projectors. (b) The scaling map of one of the projectors cut out from (a). (From [58]. © 2005 ACM, Inc. Included here by permission.)

Intuitively, the scaling map represents the pixel-wise attenuation factor needed to achieve the smooth channel display-intensity profile. The offset map represents the pixel-wise offset factor that is needed to correct for the varying black offset across the display. Together, they comprise what we call the *display smoothing maps* for each channel. From these display smoothing maps, the *projector smoothing maps* $S_{l_j}(s_j, t_j)$ and $O_{l_j}(s_j, t_j)$ for channel l of projector P_j are cut out using the geometric warp G_j as follows:

$$S_{l_j}(s_j, t_j) = S_l(G_j(s_j, t_j)); \quad O_{l_j}(s_j, t_j) = O_l(G_j(s_j, t_j)).$$

Figure 4.32 shows the scaling maps thus generated for the whole display and one projector.

Any image projected by a projector can now be corrected by applying the following three steps in succession to *every channel*.

1. The common transfer function is applied to the input image.

2. The projector smoothing maps are applied. This involves pixel-wise multiplication of the attenuation map and then addition of the offset map.

3. The inverse transfer function of the projector is applied to generate the corrected image.

In summary, the camera-based method reconstructs the spatial photo-metric variation accurately and then finds a solution to modify the photometric response in an appropriate fashion to achieve seamlessness while maintaining a high dynamic range. However, the beauty of this method is that it not only adjusts the variations in the overlap region but also corrects for intra- and inter-projector variations without treating any one of them as a special case. This is the primary difference between this method and any existing edge-blending method that addresses overlap regions only.

Implementation. The algorithm pipeline for camera-based techniques is illustrated in Figure 4.33. Reconstruction, modification, and part of the reprojection are done offline. These comprise the *calibration* step. The outputs of this step are the projector smoothing maps, the projector inverse transfer functions, and the common transfer functions for each channel. This output is then used in the *per-projector image-correction* step to correct (using Equation (4.22)) *any* image projected on the display. In this section, we discuss some practical implementation issues.

- Geometric calibration. Geometric calibration provides the following two warps: $T_{P_i \to C}$, which relates the projector coordinate space with the camera coordinate space, and $T_{C \to D}$, which relates the camera coordinate space to the display coordinate space. Any geometric-calibration algorithm that can accurately define these two warps, either linearly [20] or nonlinearly [38], can be used for this method.

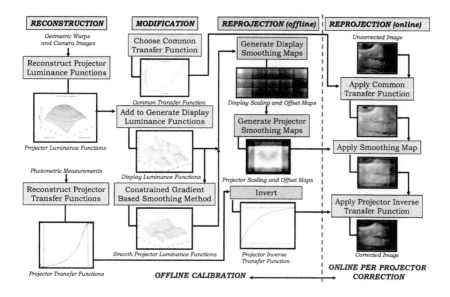

Figure 4.33. The complete algorithm. For the sake of simplicity, we have not included the black-intensity functions and the offset maps in this figure. (From [58]. © 2005 ACM, Inc. Included here by permission.)

- Photometric calibration of the camera. To linearize every camera image, the camera input transfer function (ITF) needs to be estimated. A high dynamic range imaging technique [25] is used to recover the nonlinearity of each channel of the camera, which is then used to linearize any image captured by the camera. Second, to assure that the camera does not introduce a spatial variation in addition to that which is already present in the display, its aperture is kept below f/8. Several works [57, 25] show that the spatial intensity variation of the camera is negligible in such settings.

- Generating the projector-intensity profiles. In order to generate the projector-intensity profiles, when capturing an image of projector P_i, all of the projectors that overlap with P_i are turned off. The camera images are first linearized. The image in the camera coordinate space is then converted to the projector coordinate space. To extract intensity from these images, standard linear transformations are used to convert sRGB colors to XYZ space, and then $X + Y + Z$ is computed.

  To convert to projector coordinate space, for every pixel of the projector we find the corresponding camera coordinate using $T_{P_i \to C}$ and then bilinearly interpolate the corresponding intensity from the intensity of the four nearest neighbors in the camera coordinate space.

  The projector-intensity profiles thus generated often have noise and outliers due to hot spots in the projectors and/or the camera and the nature of the screen material. The noise and the outliers are removed by a Weiner filter and a median filter, respectively. The user provides kernels for these filters by studying the frequency of the outliers and the nature of the noise.

  In most projection-based displays, adjacent projectors are overlapped to avoid rigid geometric alignment. However, the transition of intensity from the non-overlap to the overlap region is very sharp. Theoretically, to reconstruct this edge between the overlap and non-overlap regions, one would need a camera resolution of at least twice the display resolution. Given the resolution of today's display walls, this is a severe restriction. Instead, this sharp transition can be smoothed out by attenuating a few pixels at the edge of each projector. This increases the tolerance of the method to inaccuracies in reconstructing this sharp edge. This edge attenuation can be done completely in software. After generating the intensity profiles for each projector,

40–50 pixels at the edge of the projector can be attenuated using a linear function. There is no need to extract the information about the exact geometric location of the overlap regions or the geometric correspondences between projectors in this region but just an approximate idea about the width of the overlap. Further, the width of the attenuation can be changed as long as it is less than the width of the overlap region. Alternatively, a different function can be used (for example, a cosine ramp). To complement this step, we have to put the same edge attenuation back in the projected imagery. This is achieved by introducing an identical edge attenuation in the scaling and offset maps generated for each projector.

- Generating the display-intensity profiles. Following the generation of the projector-intensity profiles, the per-projector profiles are added up in the display coordinate space to generate the display-intensity profiles $\mathcal{W}_l$ and $\mathcal{B}$. For every projector pixel, we use $T_{P_i \rightarrow D}$ to find the corresponding display coordinate and then add the contribution of the intensity to the nearest four display pixels in a bilinear fashion.

Scalability. The limited resolution of the camera can affect the scalability of the reconstruction for very large displays (made up of 40–50 projectors). To design a scalable version [61] of the reconstruction method, a camera can be rotated and zoomed to estimate the intensity profiles of different parts of the display from different views [58]. These partial intensity profiles can then be stitched together to create the intensity profiles of the whole display. Figure 4.34 shows an example.

Each projector's intensity profiles are generated using different fields of view of the camera. Figure 4.35 illustrates the procedure for a display wall made up of four projectors. First, a camera is placed at position A, where the camera's field of view sees the whole display (four projectors), and and is used to perform geometric registration. The set of geometric warps from this position is denoted by G_A.

Next, the camera is moved to a new location B and/or has its zoom setting changed to get higher-resolution views of parts of the display. This movement of the camera can be performed as needed. For example, in Figure 4.35, the camera sees projectors P_1 and P_2 from B_1 and projectors P_3 and P_4 from B_2. One can then perform geometric calibration from B_1 and B_2 to get the corresponding geometric warps G_{B_1} and G_{B_2}, respectively.

Camera images are also taken for capturing the intensity profiles of each projector (similar to Figure 4.25) from B_1 and B_2. The profiles for P_1 and

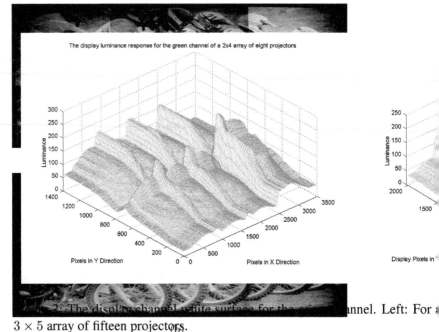

The display luminance response for the green channel of a 2x4 array of eight projectors

...2: The display channel white surface for the ...annel. Left: For a
3×5 array of fifteen projectors.
(b)

Figure 4.34. Digital photographs of a display made up of a 2×4 array of eight
projectors (4 the display is captured by the camera at different expo constrained nel whit
gradient-based intensity smoothing. This display was corrected using the scalable ure for a dis
version of the algorithm described in Section 4.4.4. (From [58]. © 2005 ACM place the can
Inc. Included here by permission.)
                                                                        sees the whol

                                                                        calibration al

                                                                        position is de

                                                                        Next we n

                                                                        to get higher

                                                                        rotate the car

                                                                        example, in I

                                                                        from B_1 and

                                                                        our geometri

                                                                        B_2 to get the

                                                                        respectively.

                                                                        We also ta

                                                                        surface for ea

                                                                        B_2. We reco

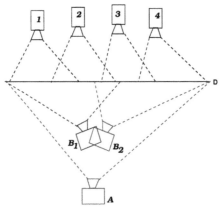

Figure 4.35. The camera and projector set-up for the scalable algorithm in Sec-from the pict
tion 4.4.4. (From [58]. © 2005 ACM, Inc. Included here by permission.) P_4 from the

                                                                        these *project*

                                                                        geometric wa

                                                                        nel white sur

                                                                        The W_g fo

However, note that both these method suffer from a serious limitation. Since the camera captures the whole display within a single field of view using a limited resolu-

P_2 are reconstructed from the images taken from B_1 using G_{B_1} and for P_3 and P_4 from the images taken from B_2 using G_{B_2}. These projector-intensity profiles are stitched together using the common geometric warp G_A from A to generate the display-intensity profiles.

Issues. One question that may occur at this moment is whether this intensity correction can introduce additional chrominance seams. Note that when photometric uniformity is applied, the shape of the intensity profile of each channel is flat after the method is applied. This assures that the proportion of red, green, and blue is identical across all pixels, and hence, no color blotches will appear.

However, the case of photometric seamlessness by the smoothing method needs greater examination. If the projector initially had no color blotches, the shape of the channel-intensity profiles given by their normalized versions will be the same. If the photometric corrections are applied to the normalized profiles, they will yield the identical different shape in three channels and, hence, not introduce any chrominance seams. However, if the projectors have initial spatial color blotches, this is essentially due to difference in the shape of the initial profiles. In such a case, when photometric smoothing is applied, the resulting shape of the profiles will be different. Thus, the color blotches cannot be removed; however, their location can change based on the way the gradient-based smoothing changed the profiles. This is illustrated in Figure 4.36

Another issue is that Equation (4.8) assumes view-independent or Lambertian displays. However, this result looks seamless for a wide range of viewing angles and distances from the wall even for a screen with gain of approximately 2.0. No artifacts are visible if the view direction makes an angle of about 20–90 degrees with the plane of the screen. Only for less

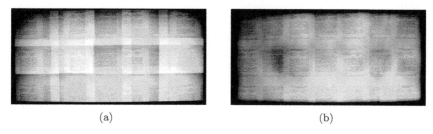

(a)   (b)

Figure 4.36. (a) Color blotches in an uncalibrated display consisting of a 5×3 array of projectors. (b) Color blotches after photometric seamlessness. (From [58]. © 2005 ACM, Inc. Included here by permission.)

than 20 degrees are the boundaries of the projectors visible, but as smooth edges, to some extent like the result of an edge-blending method.

Typically, the black offset $\mathcal{B}$ is less than 0.4% of the channel-intensity profile $\mathcal{W}_l$ and has negligible effect on the smoothing maps. An alternate version of the system assuming $\mathcal{B} = 0$ (hence, no offset map) produces very similar results except for a slight increase in the dynamic range.

4.5   Summary

This chapter analyzed the color variation in multi-projector displays extensively in Section 4.2. This analysis resulted in a classification of the variation into three categories: intra-projector, inter-projector, and overlap variation. The different parameters (such as position, projector controls, time, input, and space) that affect these variations were studied in detail. Next, a comprehensive color-variation model that captures this variation was presented in Section 4.3. Three different methods were described to color calibrate a multi-projector display by compensating for these variations in Section 4.4. These include gamut-matching methods, blending, and camera-based photometric seamlessness.

The gamut-matching methods address the inter-projector intensity and chrominance variation and require the use of a precision instrument such as a spectroradiometer. Since this method does not address overlap variation, it has to be augmented with blending techniques that address only overlap regions. Blending techniques are by far the simplest techniques since they do not need to estimate any color-model parameters and work well especially if gamut matching has been previously applied, ensuring identical intensity across different projectors.

However, these methods ignore the most significant intra-projector color variation and thus produce smoother seams in the overlap region leading to what we call a "French-door" effect. To address this problem, a more rigorous camera-based photometric seamlessness technique can be applied. This generates a display that is truly seamless in terms of intensity. The method is controlled by a single smoothing parameter that marks a trade-off between seamlessness achieved and the dynamic range sacrificed. As a result, this parameter can be set to different predefined settings for different kinds of content.

Addressing the chrominance variation in multi-projector displays is part of ongoing research. The color model for single projectors described in Section 4.3.6 has been used in a limited fashion to address the chrominance variation in single-projector displays [68] but is not covered here since it is nontrivial to extend it to multi-projector displays and, hence, requires more investigation.

All of the color-calibration methods described in Section 4.4 are content-independent, i.e., all content is corrected similarly. The camera-based photometric-seamlessness method addresses this deficiency in a limited manner by the use of the smoothing parameter. For example, for flat colors, higher smoothing parameters are required to hide all of the photometric seams, but for a moving natural scene, a lower smoothing parameter suffices. However, this method still does not consider a frame-by-frame unique compensation. Thus, the methods presented do not allow for exploiting content-dependent factors where each frame or image to be displayed can have unique color variation that can be better leveraged to achieve higher intensity or contrast. Some recent work has addressed such content-dependent corrections on single-projector displays [3].

5

PC-Cluster Rendering for Large-Scale Displays

THERE ARE TWO MAIN COMPONENTS OF a large-scale display. The first, which is the main subject of this book, is the projectors and display screen that make up the physical display. The second, and equally important, is the underlying rendering system used to generate the displayed imagery.

Less than a decade ago, this rendering component was in the domain of high-end, monolithic rendering systems, most notably Silicon Graphics machines and their equivalents. At one time, these machines were the only way to provide the necessary performance to generate the graphics output. They provided a convenient, although expensive, way to simultaneously drive multiple display outputs. Owning and operating these specialized machines, however, was no minor undertaking and required expert users for deployment, development, support, and maintenance. As a result, the reliance on these machines for rendering was another prohibitive cost in operating a tiled display.

With the recent popularization of low-cost commodity graphics cards, driven primarily by the PC gaming market, the reliance on high-end supercomputers has significantly decreased. A standard, off-the-shelf graphics card can rival the graphics throughput of the high-end machines at a fraction of the cost. This makes moving to a PC-based solution very attractive. The only caveat is that since these graphics cards are targeted for the PC market, the number of simultaneous display outputs is still typically limited to one. This means that a *cluster* of PCs is needed to control a tiled display. This chapter describes a typical cluster-rendering architecture and discusses how it can be modified to incorporate the various geometric and photometric corrections discussed in the previous chapters.

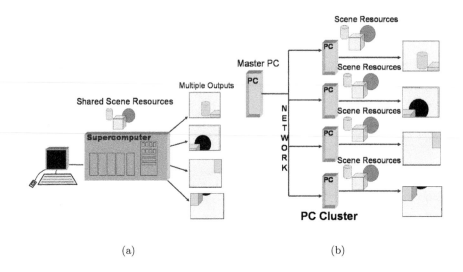

(a)   (b)

Figure 5.1. (a) Traditional tiled display rendering using a single supercomputer with multiple output. Resources can be easily shared. (b) A PC-cluster approach to rendering. Each PC has its own copy of the rendering resources (e.g., 3D geometry, texture, images, video, etc.).

5.1 Rendering Imagery for Multi-Projector Displays

One benefit offered by multi-piped supercomputers is the ability to share resources via unified memory and shared disk space (see Figure 5.1(a)). This allows 3D scene geometry, texture memory, and even disk resources to be easily shared by each processor driving a particular display output. The move to PC-cluster rendering makes this more challenging, as the data now need to be distributed.

The complexity of the underlying rendering system is related to the content to be displayed. If only a 2D image viewer, a 3D rendering system, or a video display system is needed, it is relatively easy to develop a simple client-server set-up as shown in Figure 5.1(b). In such cases, each PC in the cluster has a copy of the same resources (e.g., 3D model, 2D image, video) and is synchronized in a master-slave fashion over a dedicated network.

For more complex or generic rendering needs, several solutions have been reported that use PC clusters to drive their displays. Princeton's Scalable Display Wall [53] demonstrated one of the first viable systems using a PC cluster tied together over a high-speed network that produced imagery for a large planar display wall. Other efforts paralleled and followed this direction [37], leveraging accelerated graphics cards in PCs in lieu of expensive monolithic rendering systems. Cruz-Neira's group at Iowa State University has developed the VR Juggler [100] software, which is a comprehensive API for VR application design in such systems. Support for rendering and interaction over multiple screens driven by a cluster of workstations was added to the core distribution in 2002.

Parallel to these efforts was the development of a more general PC-based distributed rendering framework called WireGL, later named Chromium [40, 42, 43]. One benefit of Chromium is that it was designed specifically to support the OpenGL API, an accepted and established rendering API. This allows existing OpenGL applications to be supported under the Chromium framework without modification. The primary objective of Chromium is to increase rendering performance by distributing rendering tasks among a cluster of PCs, each rendering to a logical *tile*. Multiple PCs can be assigned to the same tile, partitioned tiles, or overlapping tiles. These distributed tiles are then efficiently reassembled and composited to create an output image for one or more output devices. While not explicitly targeting large-scale displays, Chromium's framework is particularly well-suited to tiled displays, where each PC tile corresponds to a projector contributing to the overall display. Not surprisingly, Chromium has today become a popular choice for use as a rendering system in tiled projector-based displays.

While shifting to a PC-cluster rendering system is instrumental in reducing rendering costs, these systems typically assume that the distributed portions of the logical display are described in a simple rectangular fashion, which does not even allow for keystone correction. This is tantamount to requiring the projector placement to be perfectly rectangular, and it restricts usage to a planar display surface, a very difficult and limiting configuration. Thus, for a practical display system, the natural step is to integrate the geometric- and photometric-correction techniques discussed in Chapters 3 and 4 into the PC-cluster rendering systems.

In the following sections, we discuss how to modify a typical distributed-cluster system to incorporate the geometric- and photometric-correction techniques that allow for casual projector alignment. In particular, we focus on the Chromium distributed rendering architecture. We use Chromium because recent versions have support for planar display arrangements that allow for nonrectangular alignment (as shown in Section 3.2.1) and therefore only require camera-based techniques to determine the projectors' homographic mapping to the display surface and the necessary photometric corrections. For more complex geometric corrections, e.g., nonlinear geometric corrections or nonparametric display surfaces, we show how to modify the existing rendering framework to incorporate these more flexible image-warping procedures. While we use Chromium as an example, the information provided should allow readers sufficient insight to make similar modifications to other cluster-based rendering systems, such as VR Juggler or existing solutions developed in-house.

5.2   Distributed Rendering using Chromium

In this section, we provide a high-level overview of the Chromium distributed rendering system supporting the OpenGL API. For more details, see [43]. Understanding Chromium is best done by first examining how basic rendering works on a single PC using the OpenGL API as shown in Figure 5.2. In a single-PC set-up, OpenGL API calls are made at runtime to an OpenGL dynamic-link library (DLL). These calls are then passed down to the underlying graphics hardware, which in turn performs the rendering. The rendered image is then sent to the output device.

Chromium works by replacing the original OpenGL DLL (`opengl32.dll` on Windows platforms) with its own DLL that supports the entire set of OpenGL commands. When OpenGL calls are made by a *client application*

Single PC - Running an OpenGL app

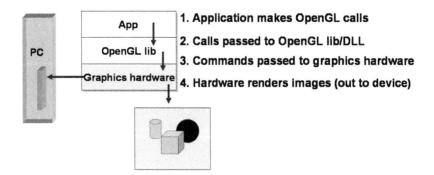

Figure 5.2. A single PC running an OpenGL application. OpenGL commands are sent to the OpenGL DLL, which passes them to the graphics hardware to render the final image.

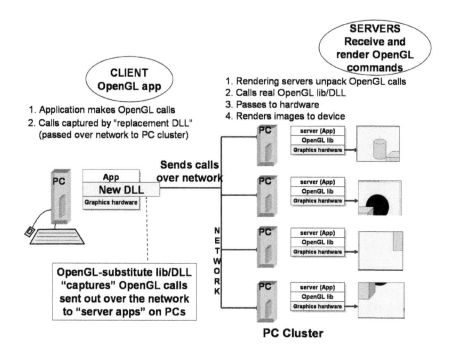

Figure 5.3. A replacement OpenGL DLL is used by Chromium [42, 43] to capture commands and pass them over the network to a cluster of PCs. These PCs have server applications waiting for incoming OpenGL commands, which are then rendered by their local graphics hardware.

to this replacement DLL, they are not passed to the graphics hardware but are instead intercepted, marshalled, and sent out over the network to a cluster of PCs. These dedicated PCs have application-level *rendering servers* that listen for OpenGL calls sent by this replacement DLL. Received calls are unpacked by these servers and passed down to the their respective hardware graphics cards, which then render the images. Figure 5.3 illustrates this set-up. Note that any mouse or keyboard events used by the application will be obtained from the PC that started the initial OpenGL application, which is now using the Chromium DLL. Conceptually, the OpenGL application started by the user is the client application, and the programs residing on the PC cluster that are rendering the imagery for this client's application are the rendering servers.

A configuration file is read in when the replacement DLL is loaded. This file specifies the network address of each PC in the cluster and information about its contribution to the large display. These individual contributions

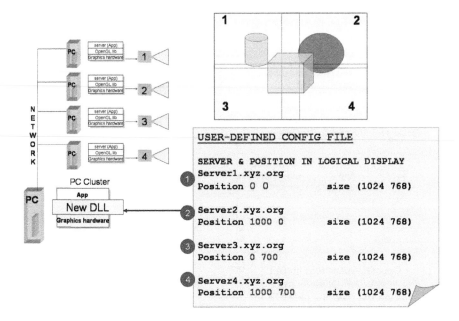

Figure 5.4. A configuration file tells the replacement DLL the network address and location of each PC in the logical display.

from each PC are referred to as *tiles*. The configuration file assumes that the virtual frame buffer resulting from the combination of all of the projectors is rectangular and starts at position $(0, 0)$. This virtual frame buffer is what we refer to as the *logical display*. Tile locations are given by their offset position with respect to the origin of this logical display and by their extent (width and height). Two or more tiles with overlapping regions will naturally result in rendered regions that correspond to the same pixels in the logical display. This ability to specify overlapping tiles is well-suited to a tiled display configuration where each PC drives a projector that contributes to a portion of the overall display as shown Figure 5.4. However, the weakness of this user-defined configuration is that the tiles need to be aligned in a precise and perfect rectangular fashion, a feat difficult to achieve manually with projectors. This particular approach also assumes that the tile in the logical display has pixels of uniform size, which is not true for projectors with keystone artifacts. Thus, the challenge lies in using a system that assumes a perfect world of rectangular tiles and achieving the geometric and photometric calibrations discussed in Chapters 3 and 4.

5.3   Seamless Tiled Displays using Chromium

As mentioned in Chapters 3 and 4, there are two main steps in generating a seamless display. The first is *camera-based calibration*. This involves a user-level application that gathers the required geometric data to geometrically register each projector into the image coordinate frame and the photometric data for blending or luminance smoothing. The second step is *correction*, where these calibration data are used to correct any imagery geometrically and photometrically in real time.

The calibration procedures that collect the geometric and photometric information need to be performed only periodically or whenever there is a significant change in the display set-up. This information is then used in the rendering system to apply the necessary corrections to any image interactively. While it is often desired to perform the geometric and photometric corrections at interactive rates, it is not necessarily critical to perform the calibration procedure in real time. Therefore, the overall calibration procedure can often take several minutes to perform in terms of information gathering from the camera as well as any necessary computation, especially in an academic setting where programming methodologies such as Matlab are used. However, this should be considered a one-time cost, as the information gathered can be used until the display has undergone a significant change.

In order to perform geometric and photometric correction within the Chromium framework, we need to modify the results of the geometric- and photometric-calibration procedures into a format usable by both the client and server application. This requires modifying Chromium's basic rendering servers to use the calibration data and correct the rendered imagery. We need to address the following two items.

- Registration and configuration-file generation. The first task is to create a new configuration file that reflects the spatial layout of the registered projectors. To accommodate Chromium's design, each projector's spatial location must be formulated as a rectangular tile.

- Server modification to perform correction. The underlying Chromium framework will have each PC render the image that corresponds to that projector's contribution to the logical display. However, this rendered image needs to be warped to accommodate the projector's orientation and the configuration of the display surface. Further, it has to be photometrically attenuated to perform blending or luminance smoothing. Chromium servers need to be modified in different ways to incorporate these necessary corrections.

5.3.1 Calibration

While it is not necessary to use the Chromium framework to obtain the geometric- and photometric-calibration data to generate the configuration file, the existing unmodified rendering framework provides a convenient way to aid in this process. This is because the calibration application can be written as an OpenGL application that generates the necessary projected fiducials for geometric registration as well as the necessary color image patterns for the photometric registration.

Assume that we have an OpenGL application running on a PC that also has access to a camera observing the display (e.g., camera attached via USB, FireWire, or an analog frame grabber). The application is running on top of the Chromium framework. A dummy configuration file specifies the network names of the PCs in the cluster with the projectors arranged in a non-overlapping configuration. This non-overlapping configuration does not represent the real projector configuration. Instead, it serves as a convenient way to guarantee that imagery can be displayed to a single projector without being displayed by any other rendering server. The client application can also read in this configuration file. The configuration file will allow the application to know how many projectors compose the display

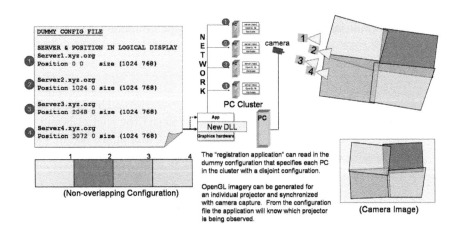

Figure 5.5. How the calibration application can be aided using the Chromium framework.

and their corresponding PC's network name. The application can then use the non-overlapping configuration of the tiles to generate imagery for each projector. Since the client application is generating the displayed imagery as well as controlling camera capture, synchronizing the display of projected patterns with camera capture for each projector can be easily done. Figure 5.5 shows an example of this type of set-up.

5.3.2 Correction

After the calibration procedure runs, the mapping $G_{(x_i,y_i)\to(s,t)}$ that maps each projector's $P_i(x_i, y_i)$ to either the display surface D or the viewer's eye, as is the case with a nonparametric surface, can be computed as described in Chapter 3. This mapping has a direct relation to the image $I(s,t)$ to be displayed. Also, the attenuation masks and the per-channel look-up table necessary for photometric correction are computed for each projector using the techniques discussed in Chapter 4.

The goal now is to modify the images rendered by the PCs using this information via Chromium. Recall that Chromium's basic design is to render imagery to rectangular tiles as shown in Figure 5.4. We can consider that these rectangular tiles generated by Chromium are sections of the $I(s,t)$ that we would like to construct on the display surface as discussed in Chapter 3. Therefore, we can consider that the Chromium framework is providing dynamic $I(s,t)$ images in real time in an interactive fashion as it runs the rendering application. The goal now is to use the computed geometric- and photometric-calibration information to modify this Chromium-generated $I(s,t)$ to produce, for each projector, the right geometric warp and photometric attenuation so that a seamless display is achieved.

Having the mapping of each projector $P_i(x_i, y_i)$ to the logical display $I(s,t)$ makes this task relatively straightforward. A rectangular tile that *bounds* each projector's mapping into the image $I(s,t)$ is computed. This tile's position and extent in the logical display is saved to create a new configuration file. This new configuration file now reflects the tile locations with respect to the projector's contributions in $I(s,t)$, as shown in Figure 5.6. We refer to these tiles as $\text{Tile}_i(s,t)$, denoting their connection to projector i and to the image $I(s,t)$

In some older releases of Chromium, tile extents were required to be of the same resolution as the rendering-cluster PCs' frame buffers. For example, if each PC graphics card is set to have a resolution of an XGA projector, then the bounding tiles must all be 1024×768 pixels in size.

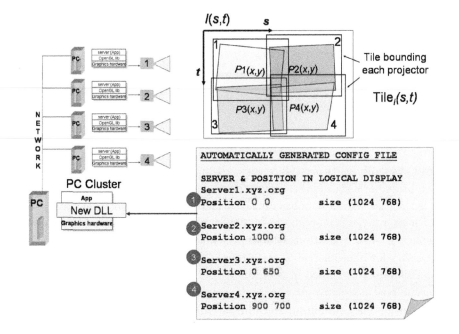

Figure 5.6. Using the Chromium spatial-configuration format, we convert our registered projector geometry into a set of equally sized rectangular tiles. This forms our logical display within the Chromium framework. The configuration file can be generated automatically.

As a result, some portions of the rectangular tile may not be used when this tile is warped based on the mapping of the projector's $P_i(x_i, y_i)$ to the $I(s,t)$. The tiles assigned in Figure 5.6 reflect this restriction, where we can see some tiles have wasted space depending on how well the tile bounds the projector's contribution. While this limitation no longer exists in new releases of Chromium that allow arbitrary tile sizes, it may still be a limitation in other cluster-rendering systems or solutions developed in-house.

After a new configuration file has been generated, it may also be necessary to update the geometric-warping information that reflects this new relationship of each projector $P_i(x_i, y_i)$ to its assigned tile $\text{Tile}_i(s,t)$ that represents a portion of $I(s,t)$. In particular, homographies between $P_i(x_i, y_i)$ and its mapping to $\text{Tile}_i(s,t)$ may need to be computed, or, if piecewise-linear techniques are used, the triangulated meshes $T^D(s,t)$ may need to be adjusted for each projector to reflect the new mapping $T_i(x_i, y_i) \rightarrow T_{\text{Tile}_i}(s,t)$. Figures 5.9 and 5.10 show examples.

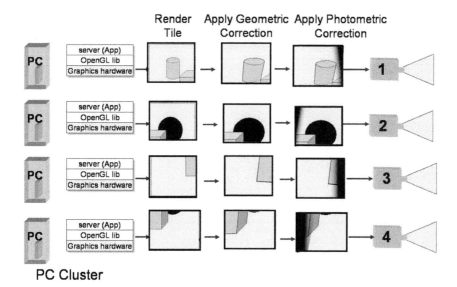

Figure 5.7. The overall pipeline for generating projected imagery.

5.3.3 Incorporating Correction

The overall pipeline for generating display imagery based on the PC cluster with incorporated image correction is depicted in Figure 5.7. The PC cluster renders the imagery in the normal fashion. These rendered images are then geometrically corrected and adjusted photometrically based on the calibration information. This correction can be implemented in a variety of ways, including special hardware placed between the PC and the projector, or even hardware placed on the projector itself.

We describe an implementation that exploits the graphics hardware used by the Chromium rendering servers. In this case, the Chromium servers are modified so that each server on start-up reads in the information necessary to correct that image. Figure 5.8 shows a diagram of the modification. Rendering servers have added "correction" routines that apply the necessary geometric warp and photometric adjustments per tile. The idea is to render the original tile, copy it to texture memory, and then apply the geometric warping and photometric attenuation routine, exploiting the graphics hardware's texture-mapping and shading capabilities.

One key question is *when* to perform the correction to the rendered image. This should be done just before the image is displayed. All cluster-rendering systems should have some barrier mechanisms that synchronize

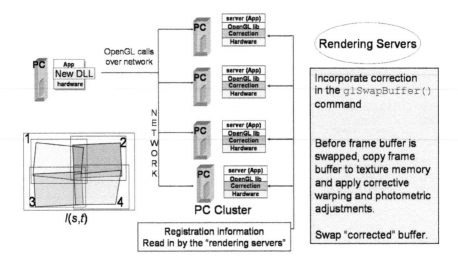

Figure 5.8. Modified Chromium diagram.

the frame display across all of the projectors. Conveniently, for OpenGL applications, this is done when the **glSwapBuffer()** command is called. Almost all OpenGL applications are double-buffered, and those that are not can be easily converted to use double-buffering. The appropriate place to perform geometric and photometric correction, therefore, is after the rendering servers receive the **glSwapBuffer()** command from the client.

Recall that the Chromium servers receive OpenGL commands from the client application. One of these commands will be the **glSwapBuffer()** command. Therefore, we can modify the existing Chromium servers' code where it implements the **glSwapBuffer()** command as follows:

```
function Chromium_glSwapBuffer() {

/* Step 1 -- Save Current State Information */
saveCurrentGLState();

/* Step 2 -- Adjust OpenGL view frustum for 2D processing */
adjustViewFor2dProcessing();

/* Step 3 -- Copy current frame buffer, i.e., rendered tile, to
texture memory */
glBindTexture(_FrameBufferTexture_);
glCopyTexSubImage2d( . . . );
```

```
/* Step 4 -- Warp the tile back to the frame buffer */
performGeometricCorrection();

/* Step 5 -- Apply Photometric Correction to frame buffer */
performPhotometricCorrection();

/* Step 6 -- Swap the buffer, i.e., display this frame buffer */
swap_buffers();

/* Step 7 -- Restore original OpenGL state */
restoreGLState();

}
```

5.4 Implementing Various Corrections in OpenGL

The geometric and photometric corrections can be implemented directly using OpenGL. The following sections give descriptions and sometimes OpenGL pseudocode for performing linear, nonlinear, and piecewise-linear geometric corrections discussed in Chapter 3. This is followed by a description of the implementation of the photometric correction discussed in Chapter 4.

5.4.1 Linear Geometric Correction

Consider Figure 5.9, where the rendered tile's relationship to the projector can be expressed as a homography. It seems reasonable that the correction could be performed using standard OpenGL texture mapping, where a 2D quadrilateral is drawn with corners coinciding with the projector's frame-buffer corners with assigned texture coordinates that correspond to the frame buffer's corners' positions in the 2D tile, which is now copied to the texture buffer. Unfortunately, this does not work in OpenGL. When the texture coordinates are assigned in OpenGL between a geometric primitive and a texture, the OpenGL API automatically computes the necessary 3×3 matrix that maps the pixels from the texture to the vertices. However, even when four points are specified, this mapping is not computed as a homography but instead as a more restrictive affine warp, which can only map triangles to triangles and not quadrilaterals to quadrilaterals as done by a homography. Thus, specifying the locations of the projector's corners in the texture image will not have the desired effect. The OpenGL API

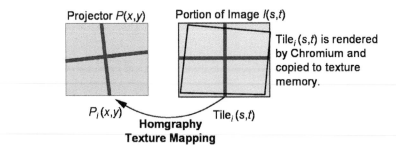

Figure 5.9. Diagram of warping procedure. The tile is rendered by standard Chromium operations. A portion of this tile is then warped to the projector's frame buffer using a homography.

does, however, allow a texture matrix, which can be a homography, to be specified by the user. For compatibility reasons, due to OpenGL's use of 4×4 matrices for 3D transforms, the homography must be specified as a 4×4 matrix instead of as a 3×3 matrix:

$$
H = \begin{pmatrix} h_1 & h_2 & 0 & h_3 \\ h_4 & h_5 & 0 & h_6 \\ 0 & 0 & 0 & 0 \\ h_7 & h_8 & 0 & 1 \end{pmatrix}.
$$

Note that this user-specified matrix is applied in addition to the texture matrix computed automatically by OpenGL. A simple workaround to get OpenGL to apply only the user-specified homography is to set up a quadrilateral whose corners map directly to the texture-map corners. This forces OpenGL to compute the texture matrix as the identity matrix. As a result, the user-specified matrix is applied directly to the texture. The pseudocode to perform this procedure is as follows:

```
/* Homography Approach */

glMatrixMode(GL_TEXTURE);
glLoadMatrix(Homography H between P(x,y) and Tile(s, t));

/* Specify a direct mapping for the four corners of the texture to
the frame buffer */
glTexCoord2d(0, 0);
glVertex2d(0, 0);

glTexCoord2d(1, 0);
glVertex2d(1, 0);
```

```
glTexCoord2d(1, 1);
glVertex2d(1, 1);
glTexCoord2d(0, 1);
glVertex2d(0, 1);
```

While it is an admittedly obfuscated procedure, we find that this is the most straightforward and direct way to specify a pre-computed homography in OpenGL.

5.4.2 Piecewise-Linear Geometric Correction

Surprisingly, the piecewise-linear approach is a bit more intuitive to implement than even the linear homography-based approaches. We simply need to warp the tile pixels from the texture to the frame buffer using the correspondences established by the triangulated mesh. Figure 5.10 shows an example of this procedure. This can be performed using the following OpenGL pseudocode:

```
/* PERFORM PIECEWISE-LINEAR WARP */
/* T are the triangles in the tessellated mesh   */
glBegin(GL_TRIANGLES);

for each triangle T
  for each vertex Vertex in triangle T
   glTexCoord2d(Tile_s, Tile_t);
   glVertex(P_x, P_y);
  end
end
glEnd();
```

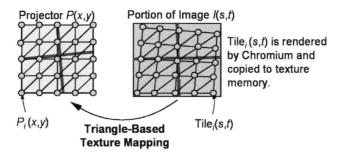

Figure 5.10. Diagram of warping procedure. The tile is rendered by standard Chromium operations. Portions of the tile are then warped to the projector's frame buffer based on the piecewise-linear mapping between the frame buffer and the tile. (© 2002 IEEE. Reprinted, with permission, from [12].)

5.4.3   Nonlinear Geometric Correction

In the nonlinear method, a nonlinear function relates the projector coordinates $P_i(x_i, y_i)$ to the image coordinates $I(s, t)$. From Chapter 3, we know that (x_i, y_i) is related to (s, t) by a polynomial. For simplicity, we denote

$$(x_i, y_i) = (f_i(s, t), g_i(s, t)), \tag{5.1}$$

where f_i and g_i are both cubic polynomials and are different for different projectors.

For nonlinear geometric correction, we use fragment-shader programs, a capability available in almost all current graphics hardware. Each projector uses a 2D look-up table to achieve the nonlinear correction. To generate this look-up table, each rendering server is given coefficients of f_i and g_i as input. The normalized tile coordinates are first converted to image coordinates. This is easily achieved from the offset and extent information in the configuration file. Then, Equation (5.1) is used to generate the projector coordinates (x_i, y_i). Finally, the generated projector coordinates are converted to the tile coordinates using the same offset and extents. During rendering, this look-up table is used in the fragment shader to texture map the tile with the appropriate pixels from the image $I(s, t)$. However, note that instead of providing the rendering servers with the nonlinear functions themselves, one can provide them with the raw look-up table in some predefined format. In that case, the calibration routine, whether using Chromium or other software such as Matlab, needs to generate the file itself. We use the functions themselves as input and generate the look-up table in the rendering server because it allows for a compact way to transfer the data from the calibration process to the correction process.

5.4.4   Photometric Correction

Here we describe how to apply camera-based photometric-correction methods using the rendering servers in Chromium. Note that this follows the geometric correction after the relevant warps have been applied to pick the appropriate parts of $I(s, t)$ for each projector.

The photometric calibration generates two important pieces of information for each projector: (a) an alpha mask A_i and (b) a 1D color look-up table h_i, for each channel encoding the inverse intensity transfer function (ITF) of the projector. The photometric correction essentially involves three steps, as explained in Section 4.4.1. These are applied to each channel independently. First, the color at each pixel is squared. This achieves the effect of the common transfer function H. Next, the transformed color is attenuated by multiplying by the value of the alpha mask at that pixel $A_i(x_i, y_i)$. Finally, the 1D lookup table h_i is applied to this transformed color to provide the final output.

We achieve this whole computation using a pixel-shader program on each rendering server. The pixel shader defines the way to manipulate the color at every pixel and hence suits our needs perfectly. The squaring and alpha-mask attenuation are easily achieved. For applying the 1D look-up table, we use 1D texture mapping. Note that for the attenuation, a multi-texturing is sufficient. However, the only way to do the other computations is via pixel shaders. Hence, we do the entire computation in pixel shaders for a simple implementation.

5.5 Demonstrations

We have built and deployed several displays using the designs outlined in this chapter. The effectiveness of these displays has been tested on a variety of arrangements. For our set-ups, we have used NVIDIA graphics cards with 32 MB, 64 MB, or 128 MB of memory. Various PC configurations have also been used. Networking is typically done with a dedicated 100 Mb ethernet switch. For the smaller set-ups, the PCs and projectors were placed on a mobile cart. This allowed us to move the entire display to temporary venues.

Figure 5.11 shows a 2×2 projector arrangement, demonstrating typical deployment of a front-projection display in a temporary venue setting. An empty room where the display is quickly assembled is shown. The projectors are arranged and registered by the camera using the calibration application, with the display being fully operational in a matter of minutes. The overall resolution of the display is approximately 2000×1293 pixels. The displayed imagery has a coverage of approximately 2×2 meters. A standard OpenGL viewing application is used to visualize 3D models. Because Chromium intercepts runtime OpenGL calls, *no* modifications to the existing OpenGL applications are needed.

Figure 5.12 shows another example set-up. In this case, three projectors were used to create a display with a resolution of approximately 3000×700 pixels. The *Atlantis* demo, often bundled with the OpenGL distribution, is shown running on the display. This display is modified by the user and re-registered in less than 60 seconds. Figure 5.12(c) and (d) show the results. The display has a spatial coverage of approximately 2.5×1 meters.

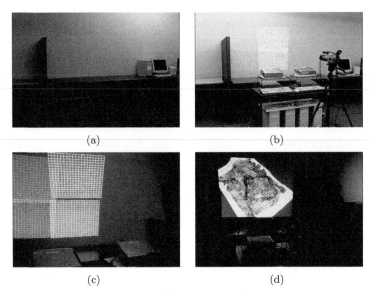

Figure 5.11. Example of the flexible large-format display being deployed. (a) An empty room. (b) How the projectors can be arranged to create a display. (c) A user adjusting the display. (d) Display in use after geometric and photometric registration.

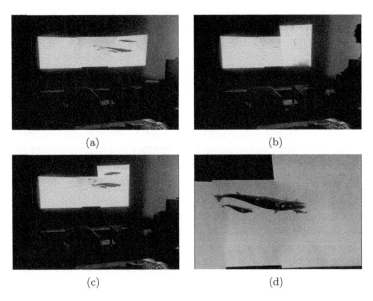

Figure 5.12. (a) Three projectors casually aligned running the *Atlantis* OpenGL demo. (b) A user modifies the display. (c) The result after re-calibration. (d) A close-up. No restrictions on projector orientation are made. (© 2002 IEEE. Reprinted, with permission, from [12].)

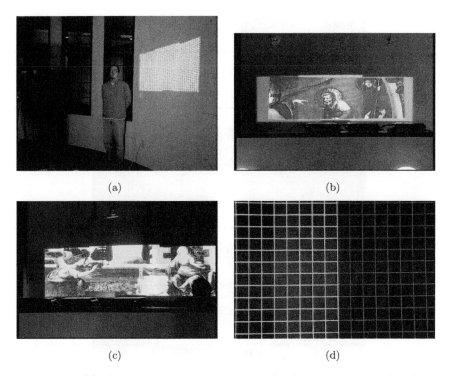

Figure 5.13. (a) Three projectors that are casually aligned on a curved surface.
(b, c) Images being displayed. (d) Geometric registration between two projectors.
(© 2002 IEEE. Reprinted, with permission, from [12].)

Figure 5.13 shows another set-up where three projectors are used, this
time projecting onto a curved surface. This demonstrates how the registration procedure can perform on a nonplanar display surface using the
piecewise-linear approaches. The resulting display has a resolution of about
2990×680 pixels and a spatial coverage of roughly 2.2×1 meters. A closeup
of the projectors with and without geometric correction is shown.

Figure 5.14 shows a three-projector array running *Quake III Arena*.
The *Quake* executable was not modified. Note how some of the pixels have
been clipped due to the nonrectangular shape of the display.

Finally, Figure 5.15 shows an interactive visualization application on
a rear-projection system made up of a 3×3 array of projectors. This
application demonstrates both the geometric and photometric corrections.
The display has a spatial coverage of about 2.5×1.8 meters and a resolution
of around 2000×3000 pixels. Note that this application can be set up
without changing the existing OpenGL program.

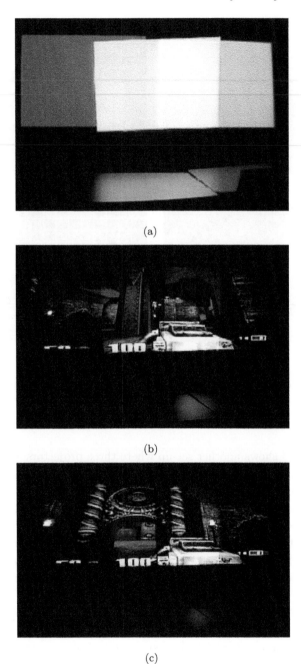

(a)

(b)

(c)

Figure 5.14. (a) Three casually aligned projectors. (b, c) The display running the *Quake III Arena* demo.

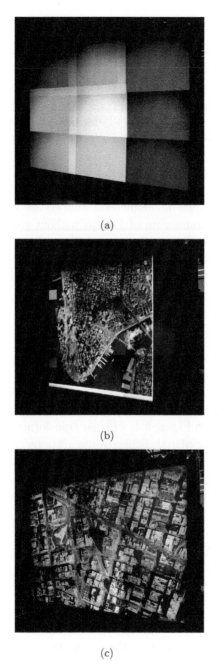

Figure 5.15. A 3 × 3 array of nine projectors using a pan-tilt-zoom image viewer. (a) The nine-projector set-up. (b) Visualization of GIS data of New York in a zoomed-out view. (c) A zoomed-in view of the same visualization.

5.5.1   Performance Issues

The OpenGL-based geometric- and photometric-correction procedure can require additional per-frame processing. However, this additional time is a constant overhead, since the exact operation is performed each time. The overhead cost is related to the performance of the graphics card. For example, with the relatively old NVIDIA GeForce3 cards with 64 MB of memory, we obtained 30 fps with geometric and photometric corrections and 60 fps without any corrections. One bottleneck is the copying of pixels from the frame buffer to texture memory. Newer NVIDIA and ATI cards allow direct rendering to texture memory, which completely alleviates this bottleneck. For example, our implementation on an NVIDIA GeForce FX 5950 Ultra shows a frame rate of 128 fps without correction and 127 fps with the geometric and photometric corrections. This can hardly be called a performance degradation. Thus, exploiting graphics hardware to realize the necessary corrections is a viable and efficient solution.

5.6   More Flexible Support of Tiles in Chromium

Since Chromium has moved to an open-source project, more flexible support for the orientation of the tiles has been incorporated. In particular, planar projective transforms (i.e., homographies) can be assigned to each tile. Now, for each entry in the display configuration, instead of an offset and extent as shown in Figure 5.4, a planar transformation can be assigned. This is very useful for planar display configurations and removes the need to modify the rendering servers to perform geometric correction, although one may still wish to incorporate changes for photometric correction.

While planar transforms are mathematically correct for aligning the imagery, small nonlinearities in the overall set-up, such as radial distortions or imperfect screens, can lead to imperfect alignment. Using the piecewise-linear approach provides a very flexible display system in such cases.

5.7   Summary

This chapter described how to use and modify a distributed PC-cluster rendering system to incorporate the geometric and photometric corrections discussed in the previous chapters. While we focus on the Chromium rendering system, the details in this chapter should provide sufficient information for the reader to modify other rendering systems, including any system developed in-house, with relatively little difficulty. This combination of camera-based geometric and photometric calibration incorporated with a PC-cluster rendering system provides a truly practical and flexible display system.

6

Advanced Distributed
Calibration

COMMON DATA-VISUALIZATION APPLICATIONS today deal with data of
incredibly large size, on the order of millions or billions of pixels. For
example, standard GIS data or satellite imagery are about 6000 × 4000
pixels, architectural drawings are about 9000 × 7000 pixels, and genome
data from the Moore Foundation's Marine Microbiology Initiative is about
10000 × 10000 pixels. In addition to scale, multi-dimensionality and multi-
modality (available in more that one form, such as text, image, video, and
sensor data) of current data make them complex and difficult to visualize. Thus, in large-scale visualization, it is critical to use a display that
can match the scale and resolution of data. Ideal for such applications is
a display that is entirely seamless, scalable, and reconfigurable and can
be easily installed. The scalability and reconfigurability help to accommodate constantly changing scale, pixel density, and aspect ratio of different
applications without burdening the user with this responsibility.

Projection technology today is ripe to provide seamless, scalable displays. Projectors are flexible display devices. They can project onto any
kind of display surface, even nonplanar, nonwhite, and non-Lambertian.
Multiple projectors can be tiled to create a display whose scale can be
changed by adding/removing projectors, pixel density can be changed by
changing the distance between the projectors and the display surface, and
form factor can be changed by changing the way the projectors are arranged in a two-dimensional array. Dramatic advancements in projection
technology have now made projectors lightweight and compact. The current pocket projector weighs about 14 ounces, can fit in the palm of your
hand, provides a 70-dpi image from a throw distance of about one foot,
uses LEDs as its lamps (increasing its longevity by an order of magnitude),

and yet is affordable at only \$700. Advances are underway to create plastic unbreakable projectors at a fraction of the cost of these pocket projectors.

In the previous chapters, we discussed several camera-based calibration techniques devised to calibrate multi-projector displays automatically, repeatedly, and inexpensively [60, 80, 103, 38, 20, 82, 19, 58, 57, 13]. However, all of these techniques have a *centralized architecture*, where one central machine/process bears the sole responsibility of achieving the geometric/color calibration by capturing specific projected patterns using a camera, analyzing them to generate the correction parameters, applying correction to different parts of the image to compensate for each projector's unique geometric and color artifacts, and finally shipping these images to the projectors to create a seamless display (see Figure 6.2). The advantage of centralized calibration is in having a common global reference frame to address the pixel geometry and color. Thus, managing multiple display units to create a seamless image is relatively easy.

However, centralized calibration is not scalable (increasing the number of projectors making up the display) or reconfigurable (changing the shape, aspect ratio, or resolution of the display). Further, it is intolerant to faults, especially in the central server. Using a single camera is one of the primary factors that limits the scalability of the calibration due to the huge mismatch between the display and camera resolutions. Using multiple cameras is also difficult in a centralized approach, and hence, most of the responsibility of managing the data and connectivity is transferred to the user, thus demanding a knowledgable user. Thus, deploying a centralized multi-projector display demands an educated user to set up the computers, projectors, and camera appropriately, input the right parameters to the central server, and maintain the whole set-up periodically.

Projectors today are affordable. Thus, building mammoth displays of a billion pixels by tiling hundreds of projectors is not unthinkable. At the other end of the spectrum, smaller, mobile, and flexible "pack-and-go" displays are very much desired for applications like map and troop-movement visualization on the battlefield. They can even be used in public venues like schools and museums. A centralized calibration architecture inhibits the realization of the full potential of using projectors in these kinds of scenarios.

Recent research has seen the development of distributed architecture and calibration techniques where no central process/computer needs to know the number of projectors, their configuration, or the geometric/photometric relationship between the projectors a priori [7]. The goal is to break away from a centralized architecture and use a distributed architecture and methodologies by which seamless high-resolution displays are built using a network of self-calibrating projector-camera systems. Such a distributed architecture is a scalable and elegant way to solve the problem of the mismatch between display and camera resolutions, since each projector-camera unit will match in terms of resolution.

This chapter explores such a *distributed architecture* and *calibration methodologies* that construct a large high-resolution display from a network of projector-camera systems. First, we describe the minimal self-sufficient unit required to realize a distributed architecture—a *plug-and-play projector* (PPP). This consists of a projector, a camera, a computation unit, and a communication unit. We then describe the architecture itself and the capabilities of a display realized by running a distributed asynchronous calibration process on this. Finally, we describe the asynchronous distributed calibration process itself. This can detect the number of neighbors of a PPP, find the position of a PPP in a large display, achieve geometric calibration and photometric blending in a distributed manner, add and remove projectors from the display dynamically for flexibility/scalability, and tolerate faults for robustness.

Distributed asynchronous calibration via plug-and-play projectors can be instrumental in using projectors for building displays that are very easy to deploy. One can imagine a user creating a display by just setting a few PPPs side by side without worrying at all about calibrating them. He can also rearrange the PPPs in a different configuration or add/remove PPPs to the existing configuration to create a display of different aspect ratio without bothering about calibration. The PPPs self-calibrate to create a seamless image in all scenarios. In fact, they can spark and foster new paradigms of interaction, especially for collaboration and visualization. Each person can carry his own plug-and-play projector since they are cost-effective and lightweight. When people meet for collaboration, their respective devices can be put together to create a seamless tiled display. Even in display walls made of a large number of PPPs, when one or more PPPs fail, the other PPPs can automatically reconfigure, recalibrate, and thus create a display of limited capability (lower resolution). In summary, distributed calibration enables *self-calibrating tiled displays*, liberating the user from any responsibility of set-up or maintenance.

6.1   System Overview

In this section, we give an overview of the system by first describing the plug-and-play projector, followed by the distributed calibration architecture, capabilities, and assumptions.

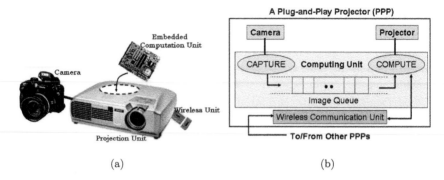

(a)   (b)

Figure 6.1. (a) A plug-and-play projector (PPP). (b) The processes on a PPP. (© 2006 IEEE. Reprinted, with permission, from [7].)

6.1.1   Plug-and-Play Projectors (PPPs)

Pixels, being the generalized purveyors of information, have been a critical commodity in any workspace with several functionalities including collaboration, visualization, and interface. The particular form of pixels provided by projectors, i.e., photons cast onto an arbitrary surface from a distance, provide a unique flexibility and mobility to this useful commodity by liberating them from the spatial constraints imposed by other displays such as CRT or LCD panels [79, 103]. However, when used in isolation, projectors act as passive digital illumination devices, pixels from which cannot provide us the desired high-end functionality, flexibility and mobility desired in a workspace. So, many researchers have proposed a marriage between projectors and cameras to provide "intelligent" pixels [82, 76, 2, 23, 21].

Consider a display unit consisting of a projector, a camera, and an embedded computing and communication hardware (see Figure 6.1). Bhasker et al. [7] call such a networked and intelligent projector a *plug-and-play projector* (PPP), and we use the same terminology. Such a PPP is self-sufficient, with the capacity to sense environmental changes (through the camera), adapt/react to those changes (through the computation unit), and share those changes with other PPPs if required (through the communication unit). One can envision the use of these units like "plug-and-play" LEGO pieces to create a scalable and reconfigurable display.

6.1.2   Architecture

In a distributed architecture, every PPP is expected to take complete control of the part of the display it is responsible for (see Figure 6.2). Thus,

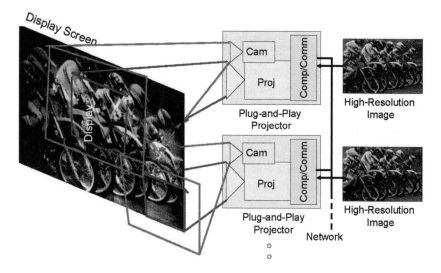

Figure 6.2. Distributed architecture for multi-projector displays. (© 2006 IEEE. Reprinted, with permission, from [7].)

it manages all of its pixels autonomously. Each PPP acts as a data client and requests the appropriate part of the data from a traditional data server (centralized or distributed) that is oblivious of the fact that the clients (the PPPs) requesting data are in reality display units. The PPPs are treated just like any other data requesting client.

Existing distributed rendering architectures such as Chromium [42, 43] and SAGE [85], discussed in Chapter 5, use distributed methodologies only for rendering the pixels and use centralized architecture for calibration and data handling. Thus, the user defines in the central server the total number of display units and the relationship of the image projected by each unit to the large image they are creating together. The centralized unit then streams the appropriate data to the dumb display units or projectors. The new architecture described here not only uses distributed rendering but also distributed calibration and data handling. This is the essential difference between these distributed rendering systems and a full-blown distributed architecture.

The second advantage of a completely distributed architecture is the asynchronous communication between different PPPs and also between the capture and projecting processes within the same PPP. This enables each PPP to start as a lone unit in the display and then discover other PPPs in the environment and their configurations. However, once this calibration phase is over, during display of application data, synchronization of the rendered frames is necessary. Since each display unit is just like a standard data-requesting client, standard distributed synchronization approaches can be used for this synchronization.

In particular, a distributed asynchronous architecture provides the following *capabilities*.

1. The PPPs calibrate themselves geometrically and photometrically to create a seamless image without any input from the user—neither the total number of PPPs making up the display and their configurations (number of rows and columns) nor the number or the identity of the neighbors each PPP has.

2. Any PPP can be added to or removed from the pool of PPPs dynamically to scale/reshape/reconfigure the display.

3. In case of faults in PPPs, the display reconfigures itself automatically to a reasonable shape.

Following are the *assumptions* about the projector-camera set-up in the PPP.

1. The camera of the PPP has a wider field of view (FOV) than the projector. Thus, the camera can see the image projected by its projector completely and can also see parts of the images projected by the neighboring PPPs. A reasonable assumption is that the camera can see about one-third of the FOV of the projectors in neighboring PPPs.

2. Since each camera captures a small part of the display, even a low-resolution (640×480) inexpensive VGA video camera suffices.

3. The camera and projector coordinate systems of a PPP and its neighbors are rotated by less than 45 degrees with respect to each other, a reasonable assumption for a rectangular array of PPPs on a planar display.

4. The camera need not be synchronized with the projector in the PPP. However, following Nyquist sampling criteria, the frame rate of the camera should be double that of the projector to assure that a pattern projected by the PPP, or an adjacent PPP, is not missed by the camera. If this criterion is not satisfied, projectors have to project their calibration patterns for more than one frame during calibration.

Since each PPP can sense changes in its neighbors via its own camera, a *camera-based communication mode* is established between adjacent PPPs via analysis of the captured images. This communication channel enables critical capabilities such as discovering local topology of the PPPs in the display array and detecting addition/removal of neighboring PPPs.

Camera-based communication can be used to communicate any information with no network overhead and hence can completely replace a network-based communication channel. However, camera-based communication is compute-intensive, so a low-bandwidth wireless communication unit on each PPP is allowed to enable network-based communication for tasks that would otherwise need complex compute-intensive image processing. Thus, both network-based and camera-based communication can be used effectively to balance the computational and network resources of the system.

6.2   Asynchronous Distributed Calibration

The asynchronous distributed calibration methodology follows a single-program multiple-data (SPMD) model in which every PPP executes the exact same program. In this section, we describe the method for a display made of a fixed number of projectors. This basic algorithm is augmented with more advanced capabilities in Section 6.3.

Each PPP runs two asynchronous processes, *capture* and *compute* (see Figure 6.1), that communicate via a shared queue of images Q and a shared boolean *CALIB*. The capture process captures images from the camera and enqueues them in Q. It only stores the images that reflect a change when compared to the last image enqueued in Q. The compute process dequeues images from Q, analyzes them, and computes different configuration and calibration parameters. It is assumed that when the compute process sends an image to be displayed on the projector, it keeps it projected unless asked to display another one. *CALIB* is used to denote the state of the PPP. A PPP can be in two states, the *calibration state*, when it calibrates itself, and the *stable state*, when it projects application data on the display. The capture and compute processes thus act like producer-consumer processes traditionally used in distributed computing environments, assuring mutual exclusion while accessing shared data structures. Algorithms 1 and 2 are the SPMD algorithms for these processes. Parts of these programs handle addition/removal of PPPs and faults (see Section 6.3).

Algorithm 1. (Pseudocode for the capture process.)

Algorithm *Process-Capture*
Image-Queue Q;
Boolean $CALIB$;
begin
1. $I = $ capture image from camera;
2. **if** $CALIB$ **then**
3.   **if** (I is different from previous image in Q) **then**
4.   enqueue I in Q;
5.   **endif**
6. **else**
7.   *Addition-Removal-Handling-for-Capture*;
8. **endif**
end

Algorithm 2. (Algorithm for the compute process.)

Algorithm *Process-Compute*
Image-Queue Q;
Boolean $CALIB$;
begin
1. $CALIB = $ True;
2. **if** ($CALIB$) **then**
3.   *Neighbor Discovery*;
4.   *Configuration Identification*;
5.   *Photometric Blending*;
6.   *Geometric Alignment*;
7. **else** // Stable state
8.   request data from server;
9.   correct and project data;
10.   *Addition-Removal-Handling-for-Compute*;
11. **endif**
end

The asynchronous distributed method starts running as soon as the projector is powered on. Each PPP starts by assuming that it is the lone display unit in the environment and is responsible for displaying the entire image. The compute process is the core of the distributed method and involves three steps to discover and then adapt to the presence of other PPPs, which together create a larger display. Figure 6.3 illustrates *the perception of each PPP* about the presence of other PPPs in the environment and its own configuration at the end of each of the following steps.

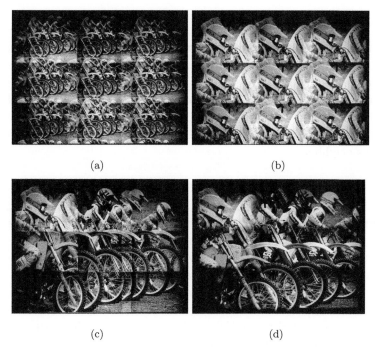

(a)   (b)

(c)   (d)

Figure 6.3. A multi-projector display made of nine PPPs. (a) At power-on. (b) The beginning of the configuration-identification process. (c) After the configuration-identification step. (d) After alignment. This is the only image that the user will see—the rest are used for illustration. (© 2006 IEEE. Reprinted, with permission, from [7].)

1. Neighbor discovery. In this stage, each PPP checks for the existence of left, right, top, and bottom neighbors using camera-based communication with adjacent PPPs.

2. Configuration identification. Next, the PPPs find the dimensions of the display array, their own coordinates in the array, and the IP addresses of all the PPPs in the display. Camera-based communication between adjacent PPPs and a network broadcast are used for this purpose.

3. Alignment. In this step, a seamless image is generated by calibrating the display geometrically and photometrically using network-based communication. For geometric alignment, a distributed homography tree technique is used. Intensity blending is used in the overlapping regions to achieve photometric seamlessness.

6.2.1 Neighbor Discovery

Neighbor discovery starts with the assumption that none of the four neighbors (left, right, top, or bottom) of a PPP exist. Each PPP projects a pattern consisting of clusters of colored blobs. The image captured by the camera of a PPP contains clusters from its own and part of the neighbor's projected pattern. The patterns are designed such that the colors and locations of the clusters can be analyzed to detect the presence/absence of the four neighbors.

The pattern. The pattern contains four *clusters*, each made of a 5×5 array of square *blobs* (see Figure 6.4). The locations of these clusters are optimized to avoid overlap of the clusters projected by adjacent PPPs and determines the maximum allowable overlap between adjacent PPPs. The particular pattern in Figure 6.4 allows a maximum overlap of 200 pixels (20% of the projector's resolution). Each cluster has a different color—red, green, blue, or white, respectively in the top-left, top-right, bottom-left, and bottom-right corners of the image. This allows a PPP to differentiate the clusters projected by itself from those projected by its neighbors and to generate a complete neighborhood description using Algorithm 3.

Identifying the location of the clusters. First, the centers of the blobs are identified using standard blob-detection techniques [80, 38]. Next, the detected blobs are grouped spatially into an array of clusters using a hierarchical agglomerative clustering approach [27] with time complexity of $O(n^2 \lceil \ln(m) \rceil)$, where n is the total number of blobs in the image and m is the dimension of each cluster (5, in this case). This method does not need as input the total number of blobs present in the image and hence can handle asynchronous display of patterns from adjacent PPPs. Pseudocode for this spatial clustering is presented in Algorithm 4.

Identifying the color of the clusters. Next, the color of each cluster is determined in the following steps.

1. The color of each blob is detected by applying chromatic filters centered at the blob centers. The detected colors, being in the camera's color space, may not coincide with the projected colors due to variations between the camera and projector color gamuts. So, each blob is assigned the projected color that has the minimum angular deviation from the detected color in the camera's RGB space.

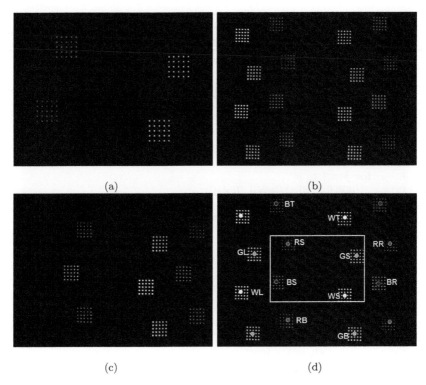

(a)   (b)

(c)   (d)

Figure 6.4. (a) The pattern projected by the projector of a PPP in the neighbor-discovery step; (b) The pattern seen by a camera of a PPP when it and all of its adjacent projectors are projecting the pattern in (a); (c) The pattern seen by a camera of a PPP when only two of its neighbors and itself are projecting the pattern. Due to the asynchronous nature of the system, there is no guarantee that all the PPPs project their pattern at the same time, so this situation is likely to occur; (d) Axis-aligned bounding rectangle enclosing the PPP's own clusters. (© 2006 IEEE. Reprinted, with permission, from [7].)

Algorithm 3. (Algorithm for discovering adjacent neighbors.)

Algorithm *Neighbor-Discovery-in-Compute*
begin
1.   project pattern;
2. $I =$ dequeue image from nonempty Q;
3.   find all clusters in I; (*Spatial Clustering*)
4.   find color, owner and centroid of each cluster;
5.   *Create* global chromatic blob tables (CBTs)
     for red, green, blue, and white;
6.   *Update* neighborhood information;
end

Algorithm 4. (Pseudocode for hierarchical agglomerative spatial clustering of blobs.)

Algorithm *Spatial-Cluster-in-Neighbor-Discovery* (*Blob*)
Input: An array *Blob* of (x, y) coordinates of the blobs.
Output: An array *Cluster* such that if *Blob*[i] and *Blob*[j]
          belong to same cluster, then $Cluster[i] = Cluster[j]$.

begin
1.   **for** $i = 1$ to n
2. $Cluster[i] = i$;
3.   **for** $j = 1$ to n
4. $d[i][j] = \text{dist}(Blob[i], Blob[j])$
5.   **endfor**
6.   **endfor**
7. $threshold = 2*\min(d)$;
8. $change = $ True;
9.   **while** *change*
10. $change = $ False;
11.   **for** $i = 1$ to n
12.   **for** $j = 1$ to n
13.   **if** $(d[i][j] < threshold)$ *AND* $(Cluster[i] > Cluster[j])$
14.   Cluster [i] = Cluster [j];
15.   change = true;
16.   **endif**
17.   **endfor**
18.   **endfor**
19. **endwhile**
end

---

2. Due to small gamut variations across the projector and camera, all of the blobs in a cluster may not be assigned the same color. Therefore, cluster colors are determined by majority voting of the colors of the component blobs.

3. Each PPP creates a *chromatic blob table* (CBT) for *each color*. The CBT lists the centers of all detected blobs in the camera coordinate along with the ID, color, and center of the cluster they belong to.

Identifying the owners of the clusters. Any cluster detected by the PPP in the previous step either belongs to itself or to its adjacent PPP. The owner of each cluster is identified using the following steps. Four connected neighbors are considered and all computations are done on the centers of clusters.

1. Since each PPP is guaranteed to see its own pattern before or along with the pattern of adjacent PPPs, an axis-aligned bounding rectangle enclosing the PPP's own clusters is deciphered (see Figure 6.4(d)). This can be done by finding the minimum x from the red and blue CBTs, the maximum x from the green and white CBTs, the maximum y from the blue and white CBTs, and the minimum y from the red and green CBTs.

2. Each cluster is assigned its closest corner in the rectangle.

3. Based on the color of the cluster and its associated rectangle corner, the ownership of the cluster is resolved. For example, the three clusters R, G, and W associated with the top-right corner in Figure 6.4(d) belong to the right neighbor, self, and top neighbor, respectively. Figure 6.4 shows the centers of the chromatically classified clusters and labels these centers to denote their ownership. The first letter of the labeling denotes the color (R, G, B, or W), and the second letter denotes the ownership (S for self, and L, R, B, or T for its four neighbors). The CBT is also updated to include the owner of each cluster. Further, for each PPP, its neighborhood information is resolved during this process.

6.2.2   Configuration Identification

In the configuration-identification step, each PPP finds the display array dimension, its own coordinate in it, and the IP addresses of all the PPPs in the display. Binary-coded bit patterns embedded in the cluster of blobs are used to convey each PPP's beliefs about where it is in the display and the total dimensions of the display. Each PPP starts by believing that it is the only node in a display of dimension 1×1. Multiple rounds of camera-based communication between adjacent projectors follow when each PPP updates its own row r and column c, and the total rows m and columns n. Update rules enable *propagation* of these parameters to all of the PPPs in the display. This results in convergence to the correct configuration at each PPP. This is followed by a network-based communication step where each PPP gathers the IP address of all other PPPs in the display. The detailed pseudocode is presented in Algorithm 5.

The pattern. The pattern for this step is derived by binary-encoding all of the clusters in each PPP in a similar fashion. Each blob denotes 1 when off and 0 when on. The first two rows are used to encode the row r and

---

Algorithm 5. (Pseudocode for the configuration-identification step.)

Algorithm *Configuration-Identification-in-Compute*
begin
1.   global $r = c = m = n = 1$;
2.   global $r_S = c_S = m_S = n_S = isdone = $ False;
3.   **if** (I am the top-left PPP) **then**
4. $r_S = c_S = $ True;
5.   **endif**
6.   **do**
7.   encode bits in grids and project ID pattern;
8. $I = $ dequeue image from non-empty Q;
9.   **if** all blobs in I present in CBTs **then**
10.   detect the IDs of the neighbor;
11.   update r, c, m, n, and the status; (*Update-IDs*)
12.   **else** // New neighbor is detected
13.   project neighbor-discovery pattern;
14.   find new grids in I and add to CBTs;
15.   update neighborhood information;
16.   reset r, c, m, and n to 1 and status bits to False;
17.   **if** (I am the top-left PPP) **then**
18. $r_S = c_S = $ True;
19.   **endif**
20.   **endif**
21. **until** *isdone*;
22. *Broadcast* MSG(IP,r,c);
23. **for** $i = 1$ to $m \times n$
24.   *Receive* Msg from non-empty Msg-Buf;
25.   *Create* and *Update* IP-Address-Table;
26. **endfor**
27. *Fault-Handling*;
end

---

column c of the PPP, and the third and fourth rows are used to encode the total number of rows m and columns n, respectively. The last row is used to denote five *status bits*—the first four bits denote if r, c, m, and n have converged to the final correct values on this PPP. The final bit *isdone* is turned on when all of the four status bits are set to denote the completion of the configuration-identification process on the PPP. Figure 6.5 shows an example of the binary encoding.

Finding the display dimension and the PPP coordinate. Configuration identification starts with each PPP setting its (r, c) and (m, n) to $(1, 1)$ and all its status bits to False. Then, each PPP performs the following steps in

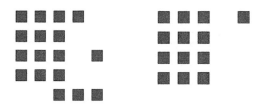

Figure 6.5. The binary-encoded grids with $(r, c) = (1, 3)$, $(m, n) = (2, 3)$, $(r_s, c_s) = (\text{True}, \text{True})$ and $(m_s, n_s, isdone) = (\text{False}, \text{False}, \text{False})$ (left) and $(r, c) = (2, 3)$, $(m, n) = (3, 3)$, $(r_s, c_s) = (\text{True}, \text{True})$ and $(m_s, n_s, isdone) = (\text{True}, \text{True}, \text{True})$ (right). (© 2006 IEEE. Reprinted, with permission, from [7].)

an iterative manner until *isdone* is set, indicating its convergence to the correct values of (r, c) and (m, n).

1. The PPPs capture the image of the encoded bit patterns and decipher the (r, c), the (m, n), and the status bits of the neighbors from the image captured by the camera. This is done by analyzing the presence or absence of blobs at the blob centers stored in the CBTs.

2. The information about its neighbors is used to update its own (r, c), (m, n), and status bits following some *update rules*. The details of this algorithm are presented in Algorithm 6.

3. The PPP then uses this information to change the embedded binary coding in its clusters and projects the new image.

The above steps on each PPP result in *propagation* of the values of the encoded parameters in the following way.

Algorithm 6. (Pseudocode for updating IDs used in configuration identification.)

Algorithm *Update-IDs-in-Configuration-Identification*
begin
1. $(r, c) = (\max(r(L), r(T) + 1), \max(c(L) + 1, c(T)))$;
2. $r_S = (T = \phi) \| r_S(L) \| r_S(T)$;
3. $c_S = (L = \phi) \| c_S(L) \| c_S(T)$;
4. $(m, n) = (\max(r, m(B), m(R)), \max(c, n(B), n(R)))$;
5. $m_S = r_S \& ((B = \phi) \| m_S(R) \| m_S(B))$;
6. $n_S = c_S \& ((R = \phi) \| n_S(R) \| n_S(B))$;
7. $isdone = m_S \& n_S \& r_S \& c_S$;
end

---

1. The PPP with no left and top neighbor (top-left PPP in the array) initiates the process and indicates that its (r, c) of $(1, 1)$ has converged by setting appropriate status bits.

2. Each PPP updates its (r, c) parameters from the *top or left* neighbor. This process continues, and the row and column changes propagate from the top-left of the display to the bottom-right in a breadth-first manner where PPPs in the same level of the tree perform the updates in parallel. This *front propagation* of (r, c) completes in $O(\ln(mn))$ steps.

3. When the bottom-right projector detects convergence of its (r, c) parameter, it sets the (m, n) to be the same as its (r, c) and turns on its *isdone* status bit to indicate convergence to the correct configuration parameters.

4. Each PPP now updates its (m, n) and *isdone* from its *bottom or right* neighbor, leading to a *back propagation* of parameters from the bottom-right to top-left of the display, again in a breadth-first manner in $O(\ln(mn))$ steps.

Thus, each PPP discovers the correct configuration parameters using only camera-based communication between adjacent projectors.

Finding IP addresses of all PPPs. Next, each PPP broadcasts its coordinates in the display along with its associated IP address over the network. On receiving this broadcast message, each PPP creates a table that maintains the coordinate of every PPP in the display along with the associated IP addresses. This step enables network communication between adjacent PPPs during alignment.

Handling race conditions. In an asynchronous system, it is possible that a neighbor of a PPP is performing its neighbor-discovery step while the PPP is in its configuration-identification step. This situation is detected by identifying appearances of new blobs in the captured image that are not present in the CBT. To handle this race condition, the PPP *aborts* its current step and goes back to the neighbor-discovery step where it lets its neighbor know of its presence. It indicates this by turning off its convergence bits, which enables propagation of this information to other nonadjacent PPPs. This also allows all the PPPs in the display to stall their convergence until information from the new PPP propagates.

Algorithm 7. (Pseudocode for geometric alignment and photometric blending.)

Algorithm *Alignment-in-Compute*
begin
1.   Root = False;
2.   **forall** *neighbors* ≠ φ **do**
3.   compute local homography to neighbor;
4.   find overlap with neighbor and apply blending;
5.   **endfor**
6.   **if** (I am the center PPP) **then**
7.   Root = True; Homogrphy-to-Root = I;
8.   **else**
9. (H, S) = *Receive* homography H and
            sender ID S from nonempty Msg-Buf;
10.   Homography-to-Root = $H \times$ Homography-to-S;
11. **endif**
12. send MSG(Homography-to-Root, myID) to all neighbors;
13. clean up Msg-Buf to delete unused homographies;
end

6.2.3 Alignment

In the alignment step, the PPPs find their exact geometric relationship with each other (amount of overlap, relative alignment of images) and use it for geometric alignment and photometric blending. First, each PPP uses the Hungarian method to detect correspondence between the blobs in the CBTs with those in the projected pattern and computes the local homography with each neighbor. This, in turn, is used to compute the overlap with its neighbors and blend it photometrically. To align the images geometrically, we use a distributed homography tree technique that aligns the images from all PPPs with respect to one reference PPP, yielding a seamless display. A detailed description of the process is presented in Algorithm 7.

Correspondence detection via Hungarian method. Previous methods establish correspondences between detected blobs in the camera space and the projected blobs in two different ways: binary-code the blobs and project them in a time-sequential manner [80, 103]; or project all blobs in one frame and then determine some distance parameters to walk along the blobs in scanline order in the projector coordinates [38, 20]—usually, additional patterns are projected to determine these distance parameters. Both methods use multiple patterns for correspondence detection. In an asynchronous distributed system, tracking multiple frames from each PPP is not viable.

So, a novel way to detect correspondences using the Hungarian method is used.

The Hungarian method is a strongly polynomial combinatorial optimization algorithm due to Kuhn [48] and later revised by Munkres [67]. It is used to solve bipartite matching (i.e., the assignment problem). The spatial clustering that generates the clusters in the neighbor-discovery stage does not impose any order to the clustered blobs. To find the order, we generate a generic template by finding the axis-aligned bounding rectangle for the detected cluster and populating it with 5×5 array of blobs. Then, a cost matrix is computed by taking the Euclidean distance between every detected-blob and template-blob pair. Each element represents the "error" induced when suggesting that particular assignment. The Hungarian algorithm then operates on this matrix to find the assignment of detected blobs to template blobs that minimizes total assignment error (the sum of the square of all distances). Thus, we order the blobs robustly and automatically. From the known order and color of the blobs, we can find the exact correspondence of blobs between the camera and projector coordinates.

The Hungarian method, however, is applicable only in scenarios where the cameras and projectors of adjacent PPPs are rotated by less than 45 degrees with respect to each other. This is a reasonable assumption for a rectangular display. However, for more general arrangements of projectors, the color coding of the clusters can be used to provide more information to this method, and relatively larger angles can also be tolerated.

Local homography calculation. Next, the correspondences are used to calculate homographies, a linear relationship tying the different device coordinates. Let the projector of a PPP be denoted by P and the camera by C. The homography between two devices A and B is denoted by $H_{A \to B}$. Each PPP calculates the *self homography* relating the PPP's own projector and camera, $H(C_{r,c} \to P_{r,c})$. It also computes the *local homography* to each neighbor, $H(P_{(r+k),(c+k)} \to P_{r,c})$, where $k \in \{-1, 0, 1\}$, as $H(P_{(r+k),(c+k)} \to C_{r,c}) * H(C_{r,c} \to P_{r,c})$.

Local photometric blending. Using the local homography, each projector finds the overlap with its neighbor and applies a linear or cosine blending (in the horizontal or vertical direction) to the RGB colors in this region [13].

Distributed geometric alignment. A distributed methodology for the homography-tree technique [20] is used to achieve geometric alignment. This starts with election of a PPP close to the center of the display as the root of the homography tree. All other PPPs align themselves with respect to the root. The homography tree is built in a breadth-first fashion in $O(\ln(mn))$ steps using network-based communication across adjacent PPPs. The process is initiated by the root. Each PPP sends its homography-to-root to all its neighbors, which augment this with their local homographies to generate their own homography-to-root. This augmentation propagation continues until all nodes have computed their homography-to-root. Note that just as in a centralized homography-tree technique, errors can accumulate along the paths from the root, creating larger errors at PPPs farther from the root. Medium-sized displays of 9–12 projectors show a maximum error of 2–3 pixels. However, due to limitations in human perception, this error is often only visible in special patterns such as grids or checkerboards.

6.3   Advanced Features

The methods described so far achieve self-calibration of the display without any user input on the number or arrangement of the PPPs. In this section, advanced capabilities that enable addition/removal of PPPs and fault tolerance are presented. These are critical features to realize truly scalable reconfigurable displays.

6.3.1   Adding and Removing Projectors

In order to handle addition and removal of PPPs to and from a calibrated display (in the stable state), the cameras in the PPPs detect addition/removal of neighboring PPPs automatically and broadcast the information to all of the existing PPPs. On receiving this broadcast, all PPPs then switch to the calibration phase to reconfigure the display (see Figure 6.6 and Algorithm 8).

Detecting addition/removal. Local homographies are used to segment the image into different regions corresponding to overlap or nonoverlap regions of the PPP itself, or of its neighbors, or empty (where no PPP projects). This segmentation can be calculated a priori during the alignment step of calibration using the homographies with the neighbors. In the stable state, simple image-processing techniques are used to detect nonblack pixels in the empty region. Similarly, all black in any of the neighbor's region detects removal.

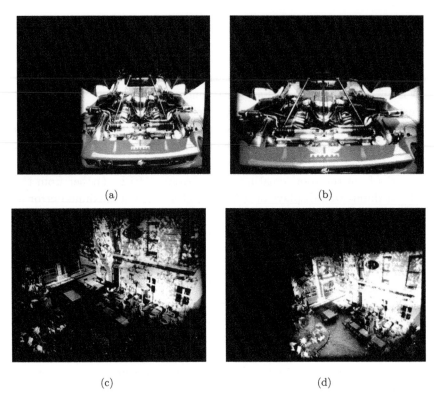

Figure 6.6. Two projectors are added to a 2 × 2 PPP display (a) to make it a 2 × 3 PPP display (b). When one projector is removed from a 3 × 3 PPP display (c), the display reshapes itself by switching off the appropriate rows and columns to generate a 2 × 2 display (d). (© 2006 IEEE. Reprinted, with permission, from [7].)

Reshaping the display. In the case of addition, a simple recalibration reshapes the display. However, since a removal creates a hole in the display, some PPPs need to be deactivated to reshape the display. For this, the message broadcast during removal contains the coordinates of the deleted PPP. Using this information, the PPPs that are on the path to the nearest vertical and/or horizontal boundary from the removed PPP deactivate themselves. The other PPPs recalibrate to reshape the display.

6.3.2   Handling Faults

One can envision hundreds of PPPs making a tiled display, especially in a public venue. In such scenarios, it is desirable to handle faults by allow-

---

Algorithm 8. (Pseudocode for handling addition/removal of PPPs in both capture and compute processes.)

Algorithm *Addition-Removal-Handling-for-Compute*
begin
1. receive Msg from non-empty Msg-Buf;
2. **if** (ADD-Msg) **then**
3. $CALIB =$ True;
4. **elseif** (DELETE-Msg)
5. $(nr, nc) =$ row and column extracted from Msg;
6.   **if** (r between nr and closest vertical boundary to nr) **or**
       (c between nc and closest horizontal boundary to nc) **then**
7.   deactivate myself;
8.   **else**
9.   *Update* (r, c, m, n) to reflect the new configuration;
10.   **endif**
11. **endif**
end

Algorithm *Addition-Removal-Handling-for-Capture*
begin
1. $I =$ dequeue image from non-empty Q;
2. process I to detect addition or removal;
3. broadcast MSG(Add/Removal, r, c);
4. $CALIB =$ True;
end

---

ing the display to run at a lower resolution even when the fault is being attended to. Following are methods to handle the most common fault of bulb outage.

If the fault occurs in the stable state, it can be handled exactly like removal of a PPP. If the fault occurs after the configuration-identification step of calibration, the following mechanisms are used to advance all the PPPs to the stable state where this is handled as a removal. Two cases occur as follows. In the first and simpler case, if the faulty PPP is not the root, all the PPPs proceed to the stable state automatically. The second case is more complicated. This happens when the faulty PPP is the root and the alignment stalls. This is handled using the invariant that Q must be empty after completion of configuration identification (since no change in patterns happens). So, the faulty root is detected by the neighboring PPPs by the existence of a nonempty Q. These PPPs broadcast a message asking everyone to advance to the stable state. On receiving this message, all PPPs comply and move to the stable state.

A fault during or before the configuration-identification step can be detected from the IP-address table in the following manner.

(a) If the top-left PPP initiating the forward propagation fails, a *conflict* results in the IP-address table with more than one PPP having $(r, c) = (1, 1)$.

(b) If any other PPP fails, it is detected as a *hole* in the IP-address table, i.e., a possible (r, c) pair is absent.

Both the conflict and the hole can be resolved by deactivating some PPPs as with the removal of PPPs in the stable state. However, in this case, instead of recalibration, the other PPPs will predict the removals and update their configuration parameters (r, c, m, n) and the IP-address table appropriately to instrument the reshaping in the subsequent alignment step.

6.4   Conclusion

There has been a plethora of work on automatic calibration of multi-projector displays in the last decade. Yet, such displays are still not commonplace, the biggest inhibition being the complexity of setting them up. A distributed calibration methodology has the potential to remove this final barrier and make multi-projector displays truly commodity products. However, very little has been done in this domain, and much needs to be accomplished before this becomes a viable technology.

Current distributed methodologies assume calibrated devices, which is a considerable hindrance for the popularity of such networks of PPPs. Closed-loop projector-camera self-calibration methodologies, both geometric and photometric, should be devised to address this issue. Current geometric algorithms assume no/little radial distortion, at least for the camera. It is difficult for commodity devices to achieve such perfect optics. Hence, homography-based methods are not really practical. At the same time, geometric alignment methods that go for a full-blown mapping usually assume a centralized single-camera technique. Thus, completely novel geometric techniques are required to achieve a high level of accuracy in a distributed architecture. The current methods also do not address the photometric nonuniformity within and across projectors beyond blending the overlap region. This needs to be addressed by designing distributed versions of the existing more rigorous photometric calibration methods such as [57]. This would involve addressing the varying dynamic range and color gamut of the sensors as well, a challenging problem by itself.

6.4.1   Moving towards Ubiquitous Pixels

Distributed calibration can have a bigger impact than just on scalable displays. *Ubiquitous pixels*—pixels anywhere and everywhere—have been envisioned by contemporary researchers as a critical component of any future workspace [26, 44]. Other critical components of future workspaces like large-scale data generation and processing, ubiquitous computing, high-performance networking, and rendering and resource management middleware have seen significant work supported by national initiatives like TeraGrid and OptIPuter [31, 45, 46]. However, ubiquitous pixels are yet to be realized by today's display technology. The key challenges are to develop methodologies to handle nonplanar, non-Lambertian, and nonwhite surfaces. We believe that distributed calibration via PPPs is the first step in that direction and has tremendous potential in realizing the vision of such ubiquitous pixels "flooding" our workspaces.

Color and Measurement

A.1 Color

Color is the response created in a sensor by a spectrum of light S. The spectrum S is a function of the wavelength λ. The function $S(\lambda)$ gives the amount of light of wavelength λ present in the spectrum.

Monochromatic color refers to light of a single wavelength. This is very difficult to achieve physically. It is known that color sensations reported by an observer with normal color vision vary as a function of the wavelength of the light stimulus. In 1976, Murch and Ball performed an experiment in which subjects were asked to identify the colors of light made up of very narrow wavelength bands (10 nm) covering the whole visual spectrum [33, 34, 28]. The observers were asked to characterize each stimulus with four numbers corresponding to the amount of blue, green, yellow, and red perceived to be present in that particular target. The results of the study showed that human perception is insensitive to wavelengths of less than 400 nm and greater than 700 nm. So 400–700 nm is called the *visual spectrum* of light. In 450–480 nm, the predominant sensation was that of blue. Green had a fairly broad band from 500–550 nm. Yellow was concentrated in a narrow band of 570–590 nm. Wavelengths above 610 nm were characterized as red. Another important observation was that most of the colors were characterized with more than two categories. For example, a 500 nm stimulus was given an average rating of 6.3 green, 2.2 blue, 0.8 yellow, and 0.1 red. The best or purest colors—defined as the maximum value estimated for one color category and the minimum value for the other three categories—indicated pure blue at about 470 nm, pure green at about

Figure A.1. The color spectrum of light for different wavelengths.

505 nm, and pure red at about 575 nm. Figure A.1 illustrates the visible light spectrum.

Achromatic color can occur only when the amount of light emitted or reflected by an object does not vary as a function of the wavelength. In other words, equal amounts of all wavelengths are present in the spectrum of the light.

Objects in the visual environment that reflect or emit distributions of wavelength in unequal amounts are said to be chromatic, which means that the spectrum of a *chromatic color* has different amounts of different wavelengths of light.

A color has two attributes associated with it: *luminance* and *chrominance*. Chrominance has two components: *hue* and *saturation*.

- Luminance. Luminance is a measure of the perceived brightness of a color and is measured in units of cd/m^2.

- Hue. Hue is determined by the relative amount of different wavelengths present in a color. For example, a spectrum as in Figure A.2

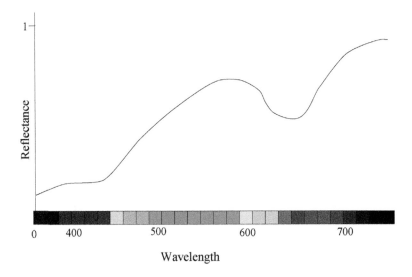

Figure A.2. The spectrum of a red color.

would have a predominantly red hue since it has more of the red wavelengths than the others.

- Saturation. Pure monochromatic light is seldom encountered in the real world. However, if a very narrow band of wavelength is taken, an observer will be able to identify a dominant hue. As the width of the band is increased, this dominant hue remains the same, but it becomes less distinct or clear. It is said then that the *hue is less saturated*. Thus, saturation depends upon the relative dominance of pure hue in a color sample. Saturation decreases (a) if the band of wavelengths in a color is increased or (b) if the amount of achromatic color added to the pure hue is increased.

A.2 Measuring Color

In order to deal with and understand color, we need to know how to measure it. The techniques to measure color comprise the science of *colorimetry*.

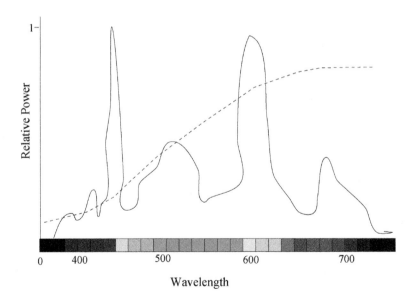

Figure A.3. Comparison of the relative spectral power distributions of a fluorescent light source (solid line) and a tungsten light source (dotted line).

A.2.1   Light Sources

The *spectral power distribution* of a light source is the power of its electro-magnetic radiation as a function of wavelength. Spectral power distribution can vary greatly for different sources of light, as shown in Figure A.3. However, note that in Figure A.3, the power values are expressed in terms of relative, not absolute, power. Such expression is sufficient for most purposes.

Measuring the color characteristics of a light source means measuring the spectral power distribution of the light source.

A.2.2   Objects

When light reaches an object, some of the light gets absorbed, and the remaining light is transmitted and reflected. The amount of light that is reflected and transmitted generally varies at different wavelengths. This variation is described in terms of *spectral reflectance* or *spectral transmittance* properties. The spectral reflectance (transmittance) of an object describes the fraction of the incident power reflected (transmitted) by the object, as a function of the wavelength. For example, the spectrum shown in Figure A.2 is the spectral reflectance of a red Cortland apple.

A.2.3   Color Stimuli

The spectral power distribution of a color stimulus is the product of the spectral power distribution of the light source and the spectral reflectance distribution of the object. The color stimulus is the result of the light that has been reflected from or transmitted through various objects. Figure A.4 shows the color stimulus that results from illuminating a Cortland apple with a fluorescent light.

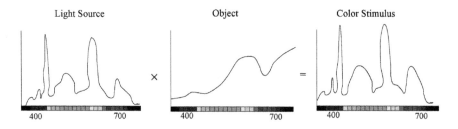

Figure A.4. Calculation of the spectral power distribution of a Cortland apple illuminated with fluorescent light.

The important thing to note here is that the perceived color of any object is not *invariant*, nor is it solely dependent on the reflectance properties of the object. An object may be made to look to be of any color by changing the light source with which it is illuminated.

A.2.4 Human Color Vision

Human color vision derives from the response of three photoreceptors, called ρ, γ, and β, respectively, contained in the retina of the eye. Each of these photoreceptors responds differently to different wavelengths of light. The approximate *spectral sensitivities* of these photoreceptors, or in other words, their relative sensitivities to light as a function of its wavelength, are shown in Figure A.5.

As mentioned in Section A.1, the human eye is insensitive to light of wavelengths greater than 700 nm or less than 400 nm. This shows that even though the color stimulus of an object suggests one color, the human perception may perceive a completely different color based on the sensitivity of the photoreceptors in the retina of the eye.

For example, Figure A.6(a) shows the spectrum of the color stimulus from an ageratum flower. From the spectrum, it appears that the color should look red, but it looks blue to the human eye. This is because the

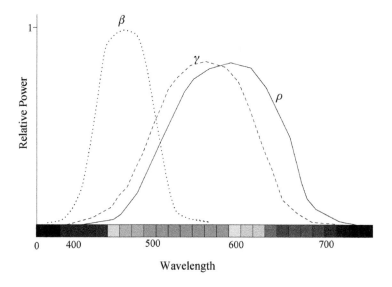

Figure A.5. Estimated spectral sensitivities of ρ, γ, and β photoreceptors of the human eye.

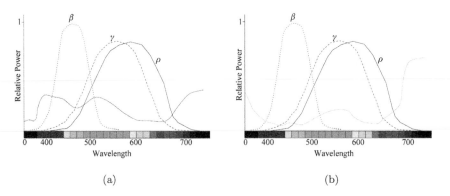

(a)   (b)

Figure A.6. (a) Color stimulus from an ageratum flower appearing blue to the human eye. (b) Color stimulus from a particular fabric sample appearing green to the human eye.

human eye is more sensitive to the blue component than the red component in this stimulus. Similarly, the stimuli produced by a fabric sample as shown in Figure A.6(b) seem green to the human observer even though the color stimulus seems to indicate otherwise.

Because of the trichromatic nature of human vision, it is very possible that two color stimuli, having different spectral power distributions, will appear identical to the human eye. This is called *metamerism*, and two such stimuli are called a *metameric pair*. In fact, metamerism is what makes color encoding possible. It is because of metamerism that there is no need to reproduce the exact spectrum of a stimulus; rather, it is sufficient to produce a stimulus that is a *visual equivalent* of the original one. Note that metamerism involves matching visual appearances of two *color stimuli* and not two *objects*. Hence, two different objects with different reflectance properties can form a metameric pair under some special lighting conditions.

A.2.5   Color Mixtures

Having considered the visual appearance of both achromatic and chromatic colors, we now consider production of colors.

Subtractive mixture of colors. The color of a surface depends on the capacity of the surface to reflect some wavelengths and absorb others. When a surface is painted with a pigment or dye, a new reflectance characteristic is developed based on the capacity of the pigment or dye to reflect and

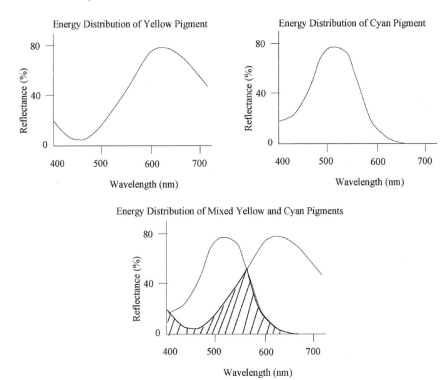

Figure A.7. Subtractive color mixture.

absorb the different wavelengths of light. Consider a surface painted with yellow pigment that reflects wavelengths in the range of 570–580 nm and another surface painted with cyan pigment that reflects wavelengths in the range of 440–540 nm. If we mix the pigments, the resulting color will be green. This is because the yellow pigment absorbs the shorter wavelengths below 500 nm and some of the middle-band wavelengths from 500–550 nm. The cyan pigment absorbs all of the longer wavelengths 560 nm and above. The energy distribution of all of these are shown in Figure A.7. Thus, the yellow absorbs the wavelengths evoking the sensation of blue, while the cyan absorbs the wavelengths evoking the sensation of yellow. Hence, what is left behind after this is a sensation of green. This is called *subtractive color mixture* because bands of wavelengths are subtracted or canceled by the combination of *light-absorbing materials*. Yellow, cyan, and magenta are the color primaries of subtractive color mixture because these are the minimal number of pigments required to produce all other colors.

Additive mixture of colors. Colors can be mixed in another fashion in which
bands of wavelengths are added to each other. This is called *additive mixture of colors*. This is also the means by which color is produced in color
displays. The surface of a color display is made up of hundreds of tiny dots
of phosphor. Phosphors are compounds that emit light when bombarded
with electrons, and the amount of light given off depends on the strength of
the electron beam. The phosphors on the screen are in groups of three, with
one phosphor emitting the longer wavelengths (red), one emitting the mid-
dle wavelengths (green), and one emitting the shorter wavelengths (blue).
The three phosphors together produce a very broad band containing all
of the visible wavelengths. Varying the intensity levels of the phosphors
produces different colors of different intensities. Thus red, green, and blue
are the primaries for the additive mixture of colors.

A.2.6   Colorimetry

CIE standard. In 1931, a special committee of the International Commis-
sion on Illumination (CIE) met to develop three standard primaries R_s, G_s,
and B_s. The major goal was to provide a numerical specification of additive color to define a set of primaries such that different amounts of the trio
produce different colors in the visual spectrum. The values R_s, G_s, and B_s
are *imaginary* primaries devised by Maxwell, which encompass the whole

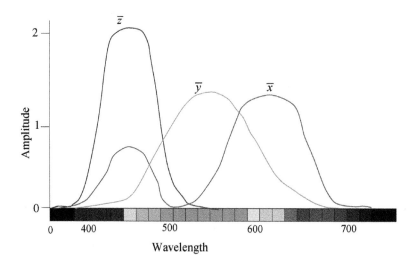

Figure A.8. A set of color-matching functions adopted by the CIE to define a
standard colorimetric observer.

visual spectrum. The amount of these primaries required to reproduce all monochromatic light can be represented as a set of *color-matching functions*, denoted by $\bar{x}(\lambda)$, $\bar{y}(\lambda)$, and $\bar{z}(\lambda)$, respectively (see Figure A.8). Note that the color-matching functions do not have any physical equivalence and are not the spectral distributions of R_s, G_s, and B_s; they are merely the auxiliary functions of how these three primaries should be mixed together to generate the metamer of the monochromatic colors.

CIE tristimulus values. It turns out that these color-matching functions are the CIE standard color primaries that model the response of the three photoreceptors in the human eye. They define what is often referred to as the *standard CIE observer*. The *CIE tristimulus values* X, Y, and Z $(X, Y, Z < 1.0)$ are the amounts of R_s, G_s, and B_s required to generate a particular color C. Let $S(\lambda)$ be the spectral power distribution of the light source. Let $R(\lambda)$ be the spectral reflectance distribution of the object. Then

$$X = k \sum_{\lambda=380}^{780} S(\lambda)R(\lambda)\bar{x}(\lambda),$$

$$Y = k \sum_{\lambda=380}^{780} S(\lambda)R(\lambda)\bar{y}(\lambda),$$

$$Z = k \sum_{\lambda=380}^{780} S(\lambda)R(\lambda)\bar{z}(\lambda),$$

where k is a normalizing factor. This is illustrated in Figure A.9.

A perfectly white object is an object that has reflectance equal to 1 throughout the visible spectrum. Also, it should be *isotropic*, i.e., it should reflect light in all directions uniformly. The factor k can be chosen in two ways. It can be chosen such that $Y = 100$ when the object is perfectly

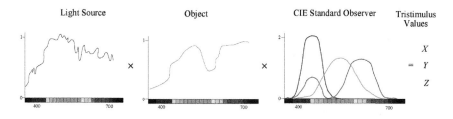

Figure A.9. Calculation of CIE tristimulus values

white; then k is called the *percent factor*. If k is chosen such that the Y value of a perfectly white object is 1.00, it is called the *factor value*.

A *colorimeter* is an instrument that can provide direct measurement of the CIE XYZ tristimulus values of a color stimulus. The value Y corresponds to *luminance*, which is the measurement of the perceived brightness. If X and Z are equal, stimuli with higher Y will look brighter.

A.2.7 Chromaticity Diagram

We define *chromaticity values* x, y, and z from the tristimulus values as follows:

$$x = \frac{X}{X+Y+Z}, \quad y = \frac{Y}{X+Y+Z}, \quad z = \frac{Z}{X+Y+Z}.$$

Notice that $x + y + z = 1$. Thus, just by knowing x and y, we cannot calculate X, Y, and Z.

Figure A.10 shows the result of plotting x and y for all visible colors. This is called the CIE chromaticity diagram and corresponds to chrominance. It has been shown that all colors with the same hue and saturation

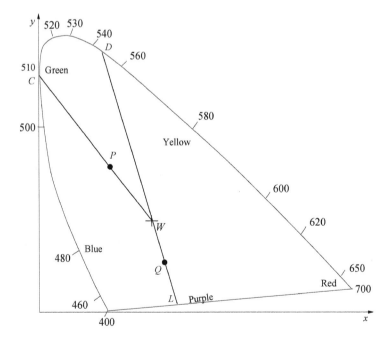

Figure A.10. CIE chromaticity diagram.

(but different luminance) correspond to the same point on the chromaticity diagram.

The interior and the boundary of the horseshoe-shaped region represent all of the visible chromaticity values. A standard white light is formally defined at W, where $x = y = z = \frac{1}{3}$. The monochromatic or the 100 percent saturated colors form the outer border of this region, while the colors in the interior of the horseshoe are unsaturated colors. The straight line at the bottom represents the various shades of purple. This diagram is very useful because it factors out the luminance and shows only the chrominance. When two colors are added together, the chrominance of the new color lies on the straight line joining the two original colors in the chromaticity diagram, the location of the new color depending on the proportion of the luminances of the two colors added.

Let P be a color in Figure A.10. The straight line joining W (white) and P meets the visual spectrum at C. Hence, P is an unsaturated form of the pure monochromatic color C. The wavelength of C is called the *dominant wavelength* of P. However, some colors do not have a dominant wavelength. For example, Q does not have a dominant wavelength since the line from W to Q when extended meets the purple line. These colors are called *nonspectral*. In such cases, the line is extended in the opposite

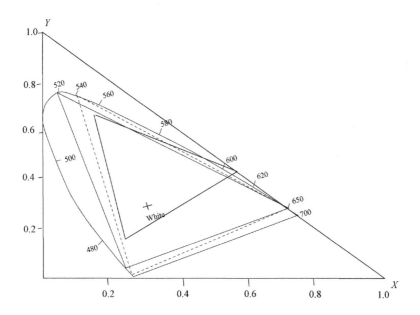

Figure A.11. CIE diagram showing three different color gamuts.

direction, and the wavelength where it meets the visual spectrum is called the *complementary wavelength* for the color. In this case, the wavelength of D is the complementary wavelength of Q. Thus, mixing D with Q will produce the achromatic color W. The *excitation purity* or *saturation* of any color possessing a dominant wavelength is defined as the ratio of the distances in the chromaticity diagram indicating how far the given color is displaced towards the spectrum color from the achromatic color. Thus, the excitation purity of P is $\frac{|WP|}{|WC|}$. For a color without a dominant wavelength, such as Q, the excitation purity is $\frac{|WQ|}{|WL|}$.

Another use of the chromaticity diagram is to specify *color gamuts*. If we draw a triangle joining any three colors in this diagram, all of the colors within the triangle can be represented by some combination of the three colors. This helps us define the *real primaries* for reproducing color, and needless to say, no set of real primaries can reproduce all of the visible colors. Figure A.11 shows three such color gamuts devised by Wright and Guild [33, 99, 34, 28]. Note that the smallest gamut cannot reproduce most of the shades of blue.

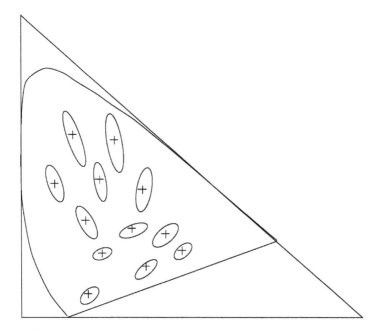

Figure A.12. Least amount of change in color required to produce a change in hue and saturation.

Advantages and disadvantages. The 1931 CIE standards have a few characteristics that render them useful for description of color in hard copy and displays. The obvious one is that the color produced by one set of primaries can be matched to the color produced by another set of primaries by adjusting them to produce the same CIE coordinates. Moreover, this models the sensitivity of a standard human observer's photoreceptors.

There are a few disadvantages of the CIE standard. The most important drawback is that the distance between two points on the CIE diagram tells nothing about the perceived color difference. In fact, from the perceptual perspective, the CIE space is nonuniform. Figure A.12 depicts a study done by MacAdam in 1942 in which the least amount of color required to produce perceivable differences in hue and saturation is shown for samples taken from all over the CIE space [28, 33, 99, 34]. Two important aspects of these data need to be noted: (a) the amount of change varies considerably for samples from different parts of the CIE space, and (b) the elliptical shape of the measurements indicates that the nonuniformity varies along the axis of the diagram. Perceptually uniform but more complicated color spaces have been designed, including the CIE Luv and Lab color spaces.

B

Perception

The end users of large high-resolution displays are humans. In order to succeed in making humans use these displays effectively, we need to know about the capabilities and limitations of human vision. This knowledge will help us in two ways. First, it will help us define the minimum requirements of such displays in terms of properties such as resolution, brightness, and so on, allowing us to build cost-effective displays. Note that we are dealing with expensive resources and that building displays with optimal capabilities only increases the total cost of the system. Second, knowing the limitations of the human visual system can help us solve certain problems easily. For example, as seen in Chapter 4, perceptual uniformity instead of strict luminance uniformity can realize displays with higher dynamic range even when using commodity projectors with severe spatial variation in luminance.

We discuss briefly the human visual system and a few relevant visual capabilities and phenomena. These are selected relevant material compiled from various references [34, 22]. Then we study the impact of these visual limitations and capabilities on several processes and choices made while building tiled displays.

B.1   Stimulus and Response

Before describing the human visual system, we will briefly visit the basic laws of perception. The most fundamental of these is *Weber's law*, which deals with the *difference threshold* [34]. The difference threshold is the minimum difference between two stimuli that a human can detect. Weber's law says that the ratio of the difference threshold and the magnitude of the stimuli is constant. This threshold is dependent on many factors, such as distance from the stimuli, ambient light, etc., and usually varies between 1 and 10 percent.

Starting from this point, Fechner showed that the magnitude of the perceived sensation is often not linearly related to the magnitude of the stimulus. Later, Stevens proved that these two are related by a power function [34]. This is *Stevens' power law* and is given by

$$\text{Perceived Sensation} = k(\text{Stimulus})^\gamma.$$

If $\gamma < 1.0$, then the change in response is more rapid than the change in stimulus. Such responses are *expansive*. For example, human response to an electric shock is expansive. On the other hand, if $\gamma > 1.0$, then the change in response is less rapid than the change in stimulus. These responses are *compressive*. Both Weber's law and Stevens' power law are true for human perception of luminance. The response of the human eye to luminance is compressive.

B.2   The Human Visual System

The human visual system comprises the sensor eye and regions of the brain, namely the lateral geniculate nucleus, the striate cortex, and the extra striate cortex regions. Figure B.1(a) shows the eye. Light enters the eye through the cornea and the lens. These two act as the focusing elements of the eye. Light is focused on the retina, which lines the back of the eye and is enriched with sensors. These sensors are activated when light falls on the retina, and they generate electrical signals that then flow through the neurons to the different regions of the brain where they are decoded and interpreted.

The iris muscles that attach the lens to the cornea of the eye are responsible for what is called the *accommodation*. When the eye is focused at different depths, these muscles lengthen or shorten the lens, letting the eye focus at different distances. Thus, these muscles help the eye to act as a lens with variable focal length. However, the eye fails to focus beyond and before certain distances, called the *depth of focus* and the *near point* of the eye, respectively.

The retina contains two types of sensors distributed in it: *rods* and *cones*. These sensors are connected to neural cells in the retina called ganglions. The ganglion cell fibers join together to form the optic nerve.

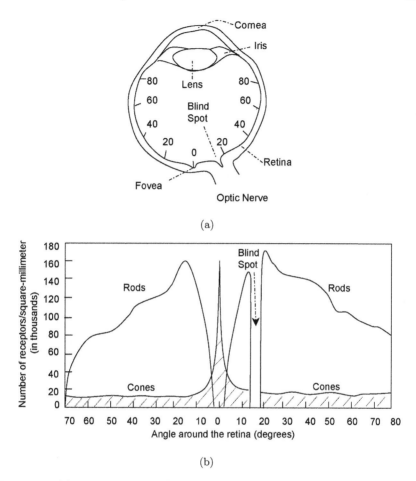

Figure B.1. (a) The human eye. (b) The distribution of the sensors on the retina, with locations on the retina in degrees relative to the fovea as indicated in (a).

The place where the optic nerve leaves the retina is devoid of any sensors and is called the *blind spot*. There is only one area on the retina that contains only cones. This area is called the *fovea*. Figure B.1(b) shows the distribution of the sensors on the retina. There are about 120 million rods, but they are distributed mostly in the periphery of the eye. There are about 5 million cones, and they are mostly concentrated near the fovea. There are three types of cones, which respond differently to different wavelengths of light and are therefore responsible for our color vision. On the other hand, there is only one type of rod, and it responds to the luminance of light and is therefore responsible for our black-and-white vision.

B.3   Visual Limitations and Capabilities

In this section, we see how the nature of the sensors and their distribution explain different visual limitations and capabilities faced in everyday life.

- Luminance and chrominance sensitivity. Just by the sheer number of the rods, it is evident that the eye is much more sensitive to luminance than to chrominance. In fact, several studies show that the eye is at least an order of magnitude more sensitive to luminance than to chrominance. This observation has been used to advantage in the design of analog color-television standards and digital image-compression algorithms such as JPEG and MPEG.

- Lower color acuity in peripheral vision. Since cones are sparsely distributed on the periphery and dense only at the fovea, our peripheral vision is mostly black and white with very low color sensitivity.

- Lower color acuity in the dark. There are a limited number of ganglion cells in the eye. These cells trigger only if the strength of the stimulus is greater than a threshold. Since there are almost 25 times more rods than cones, the number of rods attached to a ganglion cell is much higher than the number of cones. The number of sensors attached to a single ganglion cell is called its *convergence*. Just by the virtue of large numbers, the convergence of the rods is much higher than the cones. In the dark, each of the sensors receives very little light, i.e., very low stimulus. But since the convergence of the rods is high, the sum of the stimulus can easily go above the threshold and trigger the ganglion. However, due to the low convergence of the cones, they do not trigger the ganglions at such low intensity of the stimulus. Hence, we cannot see color in the dark.

- Lower resolution in the dark. The sensitivity of the eye to spatial resolution reduces in the dark for the same reason that we cannot see color in the dark. We detect a spatial resolution when two different ganglion cells are excited. Due to the higher convergence of the rods, even when different rod sensors are excited, it may mean that only one ganglion cell is triggered. However, due to low convergence of the cones, triggering a few cones is sufficient to trigger many ganglions. Since in the dark, the stimulus on the cones cannot meet the threshold required to trigger the ganglion cells, we do not see high resolution or details in the dark.

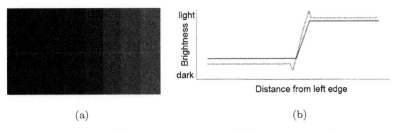

(a)   (b)

Figure B.2. (a) Mach-band effect. (b) Schematic explanation.

- Lateral inhibition. Each sensor stimulated by incoming light affects only a small region in the retina, called its *receptive field*. Rods and cones have a circular receptive field centered around themselves. Only the ganglions in the receptive field of a sensor are affected when the sensor is stimulated. However, it has been found that all of the ganglions are not affected in the same fashion. The ganglions near the center are excited, while the ones in the periphery are inhibited. This phenomenon of excitatory center and inhibitory surround is called *lateral inhibition* and creates many visual artifacts.

  The most common artifacts are *mach bands*. Figure B.2(a) illustrates the effect. The image contains eight shades of blue. Notice how the color in each of the eight strips appears to change at the transition, being darker on the left and brighter on the right. However, the change, in reality, is the same for each step. Figure B.2(b) illustrates the effect schematically. Whenever we are presented with a steep gradient as shown by the black line, we actually see the red line. This can be explained by lateral inhibition. Let us assume that the edge has brighter color on the right and darker color on the left. For the region just to the left of the edge, a large inhibitory response from the brighter right region makes it darker. On the other hand, the region just to the right of the edge receives less inhibition due to the dark surround, giving the brighter region.

- Spatial-frequency sensitivity. The density of the sensors on the retina and their convergence on the ganglion cells also limits the human ability to detect spatial frequency. In fact, our eyes act as band-pass filters, passing frequencies between two and 60 cycles per degree of angle subtended to the eye. However, we can barely detect frequencies above 30 cycles per degree. So, for all practical purposes, the cut-off frequency is 30 cycles per degree.

As expected, this band becomes narrower and narrower as the luminance decreases. Thus, in the dark, we are limited to very low-resolution vision, as mentioned above. In fact, it is interesting to note that through the evolution process, different animals have developed eyes that act as different band-pass filters depending on their requirements. For example, a falcon needs to detect its prey from high in the sky, so a falcon's eye can detect spatial frequencies between five and 80 cycles per degree. However, a goldfish does not need this kind of capability, and, hence, it can only detect low spatial frequencies in the range of two to ten cycles per degree.

- Contrast sensitivity function. A stimulus needs to have a minimum contrast so that we can decipher its spatial frequency. This is called the *contrast threshold* of that stimulus. This threshold changes depending on the spatial frequency. Sensitivity is defined as the reciprocal of the threshold. The contrast sensitivity function is shown in Figure B.3. Usually, the contrast sensitivity of humans increases with frequency until it reaches a peak and then starts to decrease, being almost zero after 60 cycles per degree. This implies that the

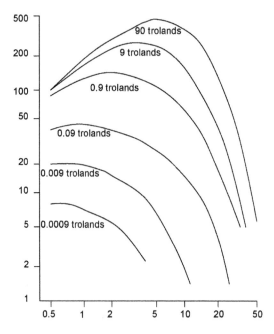

Figure B.3. The contrast sensitivity function for the human eye.

human visual system is not sensitive to smooth or low-frequency luminance changes. The sensitivity increases with more rapid changes, and for very rapid changes, the sensitivity again decreases.

Contrast sensitivity also decreases with lower luminance. This means that we need more contrast to detect different spatial frequencies at lower luminance. Also, as expected, the maximum spatial frequency we can detect reduces with decreasing luminance. Note that at very low luminance, this curve falls off almost monotonically, making the response akin to a low-pass filter.

It has been shown that our contrast sensitivity is also dependent on the orientation of the signal. We are most sensitive to horizontal or vertical signals and less sensitive to signals at oblique angles.

B.4   Relationship to Tiled Displays

Now that we have introduced the different capabilities and limitations of the human visual system, here are a few instances of how we can use this knowledge while building tiled displays.

- Resolution. Tiled displays are built with the goal of providing users with an extremely high-resolution display. But how much resolution do we really need?

  Let us assume that humans can see a maximum of 30 cycles/degree. Thus, for a person at a distance d feet from the screen, the distance of the screen that subtends an angle of one degree is given by the product of the distance (in inches) and the angle (in radians), which is $12d\pi/180$ inches. To match the maximum human sensitivity, as per the Nyquist sampling condition, we should have 60 pixels in this dimension to detect a maximum spatial frequency of 30 cycles per degree. Hence, the required resolution r in pixels/inch should satisfy

$$r > \frac{60}{\frac{12d\pi}{180}} > \frac{286}{d}.$$

  The plot of d versus r is shown in Figure B.4. Note that this is the minimum spatial-frequency requirement in horizontal and vertical directions. Our spatial-frequency sensitivities are lower for oblique directions, and, hence, we require less resolution than that given by Figure B.4. So, although theoretically we may need infinite resolution

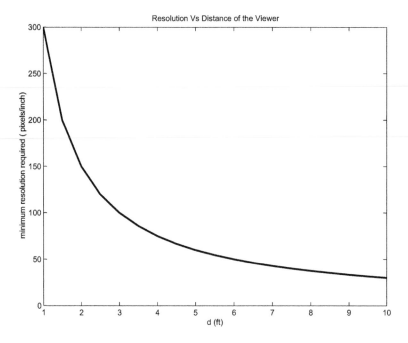

Figure B.4. The plot of the distance d of the viewer from the screen versus the minimum resolution r required by a display.

(for $d = 0$), for all practical purposes, a maximum resolution of 300 pixels/inch should be sufficient. If we make a more conservative esti-mate of the cut-off frequency as 60 cycles per degree, this maximum resolution is 500–600 pixels/inch.

- Black offset. We require higher contrast at low luminance to detect spatial frequencies. This can have an adverse effect on projection-based displays that show considerable black offset. With black offset, the contrast of low-intensity images goes through a greater percep-tual contrast reduction. Thus, the overall contrast of the display is perceived to be lower.

- Visual acuity. Due to the much higher concentration of rods in the fovea than in the periphery, we have much higher visual acuity near the fovea than in the periphery. Thus, reducing the resolution in the peripheral visual field of a tracked user may simplify the image generation without any degradation in the perceived image quality, especially when the user is at a distance and the display covers a large field of view. In fact, this increased resolution in the fovea has the perception of increasing the overall contrast of the display and is called the *punch effect*.

- Flicker. It has been shown that the flicker (temporal frequency) we can tolerate depends on the luminance of the image. We are more sensitive to flicker at higher luminance, especially in dark settings. Hence, we need higher refresh rates at higher luminance. In fact, this guides the refresh rate of our displays as they become brighter and brighter.

C

Camera Lens-Distortion Correction

Camera lenses typically exhibit some degree of nonlinearity. In some cases, such as fisheye lenses, this distortion can be quite severe. Such distortion needs to be corrected to aid in the camera-based registration techniques.

The classic lens-distortion model, proposed by Brown in 1966 [14] and later advanced by other relevant works [32], consists of two independent distortions: *radial distortion* (either barrel or pincushion) and *tangential distortion*.

Radial distortion is usually modeled by

$$r_d = r + k_1 r^2 + k_2 r^4 + k_3 r^6, \tag{C.1}$$

where r is the radial distance from the principal center in an undistorted image, r_d is the same distance in the distorted image, and k_i ($1 \le i \le 3$) are the radial-distortion parameters. Negative values of k_i result in barrel distortion, while positive values result in pincushion distortion. The principal center is a point in the image that is unchanged by radial distortion. In general, the principal center (x_c, y_c) need not be at the center of the image but is usually close to the center. Equation (C.1), when converted to Cartesian coordinates [9, 54], results in the following distortion equations:

$$x_d = x + (x - x_c)(k_1 r^2 + k_2 r^4 + k_3 r^6) = x + \rho_x, \tag{C.2}$$

$$y_d = y + (y - y_c)(k_1 r^2 + k_2 r^4 + k_3 r^6) = y + \rho_y, \tag{C.3}$$

where (x_d, y_d) and (x, y) are the distorted and undistorted image coordinates, respectively, and $r = \sqrt{(x - x_c)^2 + (y - y_c)^2}$ is the radial distance of point (x, y) from the principal center. Note that the image coordinates (x, y) have been normalized to the range $[0, 0]$ to $[1, 1]$. The terms k_i

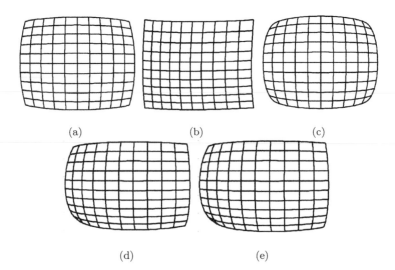

(a)   (b)   (c)

(d)   (e)

Figure C.1. Five cases of distortion.

$(1 \leq i \leq 3)$ are the radial-distortion coefficients. It is apparent from this equation that the distortion is a nonlinear function based on radial distance r from the image center. Equation (C.3) only has three distortion coefficients. This can be expanded to even more coefficients, in the form $k_i(x - x_c)r^{2i}$. However, three coefficients are often sufficient to capture most practical radial distortion.

The tangential distortion is modeled by

$$x_d = x + 2p_1xy + p_2r^2 + 2p_2x^2 = x + \tau_x, \quad (C.4)$$

$$y_d = y + 2p_2xy + p_1r^2 + 2p_1y^2 = y + \tau_y, \quad (C.5)$$

where $r = \sqrt{(x - x_c)^2 + (y - y_c)^2}$ and p_1 and p_2 are the tangential-distortion parameters.

Radial and tangential distortion are independent of each other, and, hence, their effects can be combined into a comprehensive lens-distortion equation:

$$(x_d, y_d) = (x + \rho_x + \tau_x, y + \rho_y + \tau_y). \quad (C.6)$$

Figure C.1 shows some examples of distorted grids resulting from the lens-distortion model defined by Equation (C.6). In Figure C.1(a), $(x_c, y_c) = (0.5, 0.5)$, $(k_1, k_2, k_3) = (-0.35, 0, 0)$, and $(p_1, p_2) = (0, 0)$ (only radial distortion with the principal center coincident with the center of the image). In Figure C.1(b), $(x_c, y_c) = (0.5, 0.5)$, $(k_1, k_2, k_3) = (0, 0, 0)$, and

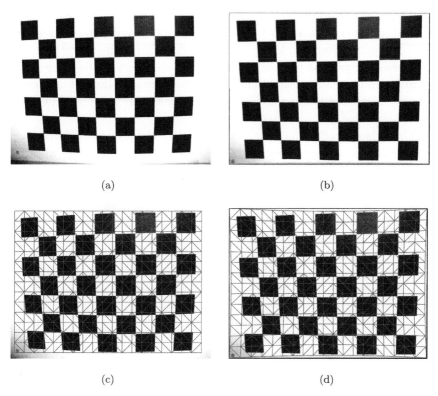

(a)   (b)

(c)   (d)

Figure C.2. (a) Image with radial distortion. (b) Corrected image. (c, d) Warping from (a) to (b) can be performed by piecewise texture mapping.

$(p_1, p_2) = (0.1, 0.1)$ (only tangential distortion with principal center coincident with the center of the image). In Figure C.1(c), $(x_c, y_c) = (0.5, 0.5)$, $(k_1, k_2, k_3) = (-0.35, -0.35, 0)$, and $(p_1, p_2) = (0, 0)$ (more higher-order radial-distortion terms added to Figure C.1(a)). In Figure C.1(d), $(x_c, y_c) = (0.6, 0.55)$, $(k_1, k_2, k_3) = (-0.35, -0.13, 0)$, and $(p_1, p_2) = (0.05, 0.05)$ (both radial and tangential distortion with principal center not coincident with the center of the image). In Figure C.1(e), $(x_c, y_c) = (0.6, 0.55)$, $(k_1, k_2, k_3) = (-0.35, -0.13, -0.016)$, and $(p_1, p_2) = (0.05, 0.05)$ (more higher-order radial-distortion terms added to Figure C.1(d)).

One practical way to solve for the coefficients k_i and p_i is to image a pattern of known features that lie on straight lines. For example, in Figure C.2, a checkerboard pattern is observed, where the checkerboard corners are known to lie on lines running horizontally and vertically. A cost function can be constructed that computes the errors of (x_d, y_d) to lines fitted to these points. The parameters k_i and p_i can then be found by minimizing the cost function via iterative energy-minimization approaches such as gradient descent. A very effective radial-distortion calibration toolbox, widely known as the Caltech Camera Calibration Toolbox, provides a user-friendly interface and code to achieve such lens-distortion correction [9]. Note that tangential distortion is significantly less prevalent than radial distortion and can often be omitted.

Once these coefficients have been computed, images captured from the camera can be corrected based on the derived k_i and p_i. Note that in some cases, this correction only needs to be applied to the extracted image features. For example, in the geometric-registration approaches, radial-distortion correction can be applied to the image features after they have been extracted and not the entire image. When it is necessary to warp the entire image, a simple tesselated mesh of regularly sampled (x, y) mapped to their corresponding (x_d, y_d) can be used with standard texture-mapping routines to obtain the warped image, as shown in Figure C.2(c) and (d).

Bibliography

[1] L. Arnstein, C. Y. Hung, R. Franza, Q. H. Zhou, G. Boriello, S. Consolvo, and J. Su. "Labscape: A Smart Environment for the Cell Biology Laboratory." *IEEE Pervasive Computing* 1:3 (2002), 13–21.

[2] M. Ashdown, M. Flagg, R. Sukthankar, and J. M. Rehg. "A Flexible Projector-Camera System for Multi-Planar Displays." In *Proceedings of Computer Vision and Pattern Recognition (CVPR)*, pp. II–165–II–172. Washington, D.C.: IEEE Computer Society, 2004.

[3] M. Ashdown, T. Okabe, I. Sato, and Y. Sato. "Robust Content-Dependent Photometric Projector Compensation." In *IEEE CVPR Workshop on Projector Camera Systems*, p. 6. Washington, D.C.: IEEE Computer Society, 2006.

[4] M. Bern and D. Eppstein. "Optimized Color Gamuts for Tiled Displays." In *19th ACM Symposium on Computational Geometry*, pp. 274–281. New York: ACM Press, 2003.

[5] R. S. Berns, M. E. Gorzynski, and R. J. Motta. "CRT Colorimetry, Part II: Metrology." *Color Research and Application* 18:5 (1993), 315–325.

[6] R. S. Berns, R. J. Motta, and M. E. Gorzynski. "CRT Colorimetry, Part I: Theory and Practice." *Color Research and Application* 18:5 (1993), 299–314.

[7] E. S. Bhasker, P. Sinha, and A. Majumder. "Asynchronous Distributed Calibration for Scalable Reconfigurable Multi-Projector Displays." *IEEE Transections of Visualization and Computer Graphics (TVCG)* 12:5 (2006), 1101–1108.

[8] O. Bimber and R. Raskar. *Spatial Augmented Reality: Merging Real and Virtual Worlds*. Wellesley, MA: A K Peters, 2005.

[9] J. Y. Bouguet. *Camera Calibration Toolbox for Matlab*. Available online (http://www.vision.caltech.edu/bouguetj/calib_doc/index.html), 2001.

[10] D. H. Brainard. "Calibration of a Computer Controlled Color Monitor." *Color Research and Applications* 14:1 (1989), 23–34.

[11] R. A. Brooks. "The Intelligent Room Project." In *Proceedings of the Second International Cognitive Technology Conference*, p. 271. Washington, D.C.: IEEE Computer Society, 1997.

[12] M. S. Brown and W. B. Seales. "A Practical and Flexible Tiled Display System." In *Proceedings of IEEE Pacific Graphics*, p. 194. Washington, D.C.: IEEE Computer Society, 2002.

[13] M. S. Brown, A. Majumder, and R. Yang. "Camera-Based Calibration Techniques for Seamless Multi-Projector Displays." *IEEE Transactions on Visualization and Computer Graphics* 11:2 (2005), 193–206.

[14] D. C. Brown. "Decentering Distortion of Lenses." *Photometric Engineering* 32:3 (1966), 444–462.

[15] B. Brumitt, B. Meyers, J. Krumm, A. Kern, and S. Shafer. "EasyLiving: Technologies for Intelligent Environments." In *Handheld and Ubiquitous Computing, Lecture Notes in Computer Science 1927*, pp. 12–29. Berlin: Springer-Verlag, 2000.

[16] I. Buck, G. Humphreys, and P. Hanrahan. "Tracking Graphics State for Networked Rendering." In *Proceedings of Eurographics/SIGGRAPH Workshop on Graphics Hardware*, pp. 87–95. New York: ACM Press, 2000.

[17] P. J. Burt and E. H. Adelson. "A Multiresolution Spline with Application to Image Mosaics." *ACM Transactions on Graphics (TOG)* 2:4 (1983), 217–236.

[18] C. J. Chen and Mike Johnson. "Fundamentals of Scalable High-Resolution Seamlessly Tiled Projection System." In *Proceedings of SPIE Projection Displays VII*, pp. 67–74. Bellingham, WA: SPIE, 2001.

[19] Y. Chen, D. W. Clark, A. Finkelstein, T. Housel, and K. Li. "Automatic Alignment Of High-Resolution Multi-Projector Displays using an Un-Calibrated Camera." In *Proceedings of IEEE Visualization*, pp. 125–130. Washington, D.C.: IEEE Computer Society, 2000.

[20] H. Chen, R. Sukthankar, G. Wallace, and K. Li. "Scalable Alignment of Large-Format Multi-Projector Displays Using Camera Homography Trees." pp. 339–346. Washington, D.C.: IEEE Computer Society, 2002.

[21] L. Childers, T. Disz, R. Olson, M. E. Papka, R. Stevens, and T. Udeshi. "Access Grid: Immersive Group-to-Group Collaborative Visualization." In *Proceedings of Immersive Projection Technology*, 2000.

[22] R. A. Chorley and J. Laylock. "Human Factor Consideration for the Interface between Electro-Optical Display and the Human Visual System." In *Displays*, 1981.

[23] D. Cotting, R. Ziegler, M. Gross, and H. Fuchs. "Adaptive Instant Displays: Continuously Calibrated Projections Using Per-Pixel Light Control." In *Proc. of Eurographics*, pp. 705–714. Aire-la-Ville, Switzerland: Eurographics Association, 2005.

[24] C. Cruz-Neira, D. J. Sandin, and T. A. Defanti. "Surround-Screen Projection-Based Virtual Reality: The Design and Implementation of the CAVE." In *Proceedings of SIGGRAPH 93, Computer Graphics Proceedings, Annual Conference Series*, pp. 135–142. New York: ACM Press, 1993.

[25] P. E. Debevec and J. Malik. "Recovering High Dynamic Range Radiance Maps from Photographs." In *Proceedings of SIGGRAPH 97, Computer Graphics Proceedings, Annual Conference Series*, pp. 369–378. Reading, MA: Addison-Wesley, 1997.

[26] T. L. Disz, M. E. Papka, and R. Stevens. "UbiWorld: An Environment Integrating Virtual Reality." In *Heterogeneous Computing Workshop*, pp. 46–57. Washington, D.C.: IEEE Computer Society, 1997.

[27] R. O. Duda and P. E. Hart. *Pattern Classification and Scene Analysis.* New York: John Wiley and Sons, 1973.

[28] H. J. Durrett, editor. *Color and the Computer.* San Diego, CA: Academic Press, 1987.

[29] O. Faugeras. *Three-Dimensional Computer Vision: A Geometric Viewpoint.* Cambridge, MA: MIT Press, 1993.

[30] J. D. Foley, A. van Dam, S. K. Feiner, and J. F. Hughes. *Computer Graphics: Principles and Practice.* Reading, MA: Addison Wesley, 1990.

[31] I. Foster and C. Kesselman. *The Grid: Blueprint for a New Computing Infrastructure.* San Francisco, CA: Morgan Kaufmann Publishers, 1998.

[32] J. G. Fryer and D. C. Brown. "Lens Distortion for Close-Range Photogrammetry." *Photogrammetric Engineering and Remote Sensing* 52:1 (1986), 51–58.

[33] E. J. Giorgianni and T. E. Madden. *Digital Color Management: Encoding Solutions.* Reading, MA: Addison Wesley, 1998.

[34] E. B. Goldstein. *Sensation and Perception (6th Edition).* Pacific Grove, CA: Wadsworth-Thomson Learning, 2002.

[35] R. C. Gonzalez and R. E. Woods. *Digital Image Processing.* Reading, MA: Addison Wesley, 1992.

[36] M. Harville, B. Culbertson, I. Sobel, D. Gelb, A. Fitzhugh, and D. Tanguay. "Practical Methods for Geometric and Photometric Correction of Tiled Projector Displays on Curved Surfaces." In *IEEE International Workshop on Projector-Camera Systems*, p. 5. Washington, D.C., 2006.

[37] M. Hereld, I. Judson, and R. Stevens. "Introduction to Building Projection-Based Tiled Display Systems." *IEEE Computer Graphics and Applications* 20:4 (2000), 22–28.

[38] M. Hereld, I. R. Judson, and R. Stevens. "DottyToto: A Measurement Engine for Aligning Multi-Projector Display Systems." Technical Report ANL/MCS-P958-0502, Argonne National Laboratory, 2002.

[39] M. Hereld. "Local Methods for Measuring Tiled Display Alignment." In *Proceedings of IEEE International Workshop on Projector-Camera Systems.* Washington, D.C.: IEEE Computer Society, 2003.

[40] G. Humphreys and P. Hanrahan. "A Distributed Graphics System for Large Tiled Displays." In *Proceedings of IEEE Visualization*, pp. 215–223. Washington, D.C.: IEEE Computer Society, 1999.

[41] G. Humphreys, I. Buck, M. Eldridge, and P. Hanrahan. "Distributed Rendering for Scalable Displays." In *Proceedings of IEEE Supercomputing*, p. 30. Washington, D.C.: IEEE Computer Society, 2000.

[42] G. Humphreys, M. Eldridge, I. Buck, G. Stoll, M. Everett, and P. Hanrahan. "WireGL: A Scalable Graphics System for Clusters." In *Proceedings of SIGGRAPH 2001, Computer Graphics Proceedings, Annual Conference Series*, pp. 129–140. New York: ACM Press, 2001.

[43] G. Humphreys, M. Houston, R. Ng, R. Frank, S. Ahem, P. Kirchner, and J. Klosowski. "Chromium: A Stream Processing Framework for Interactive Rendering on Clusters." *ACM Transactions of Graphics* 21:3 (2002), 693–702.

[44] B. Johanson, A. Fox, and T. Winograd. "The Interactive Workspaces Project: Experience with Ubiquitous Computing Rooms." *IEEE Pervasive Computing* 1:2 (2002), 67–74.

[45] Y.-S. Kee, D. Logothetis, R. Huang, H. Casanova, and A. A. Chien. "Efficient Resource Description and High Quality Selection for Virtual Grids." In *Proceedings of the IEEE Conference on Cluster Computing and the Grid (CCGrid)*, pp. 598–606. Washington, D.C.: IEEE Computer Society, 2005.

[46] G. M. Kent, J. Orcutt, L. Smarr, J. Leigh, A. Nayak, D. Kilb, L. Renambot, S. Venkataraman, T. DeFanti, Y. Fialko, P. Papadopoulos, G. Hidley, D. Hutches, and M. Brown. "The OptIPuter: A New Approach to Volume Visualization of Large Seismic Datasets." In *Ocean Technology Conference*, 2004.

[47] M. Kratz. *Access Grid (AG) Collaboration.* Available online (http://health.internet2.edu/contacts/contacts.html), 2003.

[48] H. W. Kuhn. "The Hungarian Method for Solving the Assignment Problem." *Naval Research Logistics Quarterly* 2 (1955), 83–97.

[49] J. Lai, A. Levas, P. Chou, C. Pinhanez, and M. Viveros. "BlueSpace: Personalizing Workspace through Awareness and Adaptability." *International Journal of Human Computer Studies* 57:5 (2002), 415–428.

[50] E. H. Land and J. J. McCann. "Lightness and Retinex Theory." *Journal of Optical Society of America* 61:1 (1971), 1–11.

[51] E. H. Land. "The Retinex." *American Scientist* 52:2 (1964), 247–264.

[52] K. Li and Y. Chen. "Optical Blending for Multi-Projector Display Wall System." In *Proceedings of the 12th Lasers and Electro-Optics Society 1999 Annual Meeting*, pp. 281–282. Washington, D.C.: IEEE Computer Society, 1999.

[53] K. Li, H. Chen, Y. Chen, D. W. Clark, P. Cook, S. Damianakis, G. Essl, A. Finkelstein, T. Funkhouser, A. Klein, Z. Liu, E. Praun, R. Samanta,

B. Shedd, J. P. Singh, G. Tzanetakis, and J. Zheng. "Early Experiences and Challenges in Building and using a Scalable Display Wall System." *IEEE Computer Graphics and Applications* 20:4 (2000), 671–680.

[54] L. Ma, Y. Chen, and K. L. Moore. "A New Analytical Radial Distortion Model for Camera Calibration." *CoRR* cs.CV/0307046.

[55] A. Majumder and M. Gopi. "Modeling Color Properties of Tiled Displays." *Computer Graphics Forum* 24:2 (2005), 149–163.

[56] A. Majumder and R. Stevens. "LAM: Luminance Attenuation Map for Photometric Uniformity in Projection Based Displays." In *Proceedings of ACM Virtual Reality and Software Technology*, pp. 147–154. New York: ACM Press, 2002.

[57] A. Majumder and R. Stevens. "Color Nonuniformity in Projection-Based Displays: Analysis and Solutions." *IEEE Transactions on Visualization and Computer Graphics* 10:2 (2003), 177–188.

[58] A. Majumder and R. Stevens. "Perceptual Photometric Seamlessness in Projection-Based Tiled Displays." *ACM Transactions on Graphics* 24:1 (2005), 118–139.

[59] A. Majumder and G. Welch. "Computer Graphics Optique: Optical Superposition of Projected Computer Graphics." In *Proceedings of Joint Eurographics Workshop on Virtual Environments and Immersive Projection Technology*. Vienna: Springer Wien, 2001.

[60] A. Majumder, Z. He, H. Towles, and G. Welch. "Achieving Color Uniformity Across Multi-Projector Displays." In *Proceedings of IEEE Visualization*, pp. 117–124, 546. Washington, D.C.: IEEE Computer Society, 2000.

[61] A. Majumder, D. Jones, M. McCrory, M. E. Papke, and R. Stevens. "Using a Camera to Capture and Correct Spatial Photometric Variation in Multi-Projector Displays." In *Proceedings of IEEE International Workshop on Projector-Camera Systems*. Washington, D.C.: IEEE Computer Society, 2003.

[62] A. Majumder. "Properties of Color Variation Across Multi-Projector Displays." In *Proceedings of SID Eurodisplay*, pp. 807–810. San Jose, CA: Society for Information Display, 2002.

[63] A. Majumder. "Contrast Enhancement of Multi-Displays Using Human Contrast Sensitivity." In *Proceedings of IEEE International Conference on Computer Vision and Pattern Recognition (CVPR)*, pp. 377–382. Washington, D.C.: IEEE Computer Society, 2005.

[64] T. L. Martzall. "Simultaneous Raster and Calligraphic CRT Projection System for Flight Simulation." In *SPIE Proceedings, Electroluminescent Materials, Devices, and Large-Screen Displays*, pp. 292–299. Bellingham, WA: SPIE, 1993.

[65] S. Molnar, M. Cox, D. Ellsworth, and H. Fuchs. "A Sorting Classification of Parallel Rendering." *IEEE Computer Graphics and Algorithms* 14:4 (1994), 23–32.

[66] J. Montrym, D. Baum, D. Dignam, and C. Migdal. "InfiniteReality: A Real-Time Graphics Systems." In *Proceedings of SIGGRAPH 97, Computer Graphics Proceedings, Annual Conference Series*, pp. 293–302. Reading, MA: Addison Wesley, 1997.

[67] J. Munkres. "Algorithms for the Assignment and Transportation Problems." *Journal of SIAM* 5 (1957), 32–38.

[68] S. K. Nayar, H. Peri, M. D. Grossberg, and P. N. Belhumeur. "A Projection System with Radiometric Compensation for Screen Imperfections." In *IEEE International Workshop on Projector-Camera Systems*. Washington, D.C.: IEEE Computer Society, 2003.

[69] L. O'Callaghan, N. Mishra, A. Mayerson, S. Guha, and R. Motwani. "Streaming-Data Algorithms for High-Quality Clustering." In *IEEE International Conference on Data Engineering*, p. 685. Washington, D.C.: IEEE Computer Society, 2002.

[70] *OpenCV Library*. Available online (http://opencvlibrary.sourceforge.net/), 2005.

[71] J. Owens, W. Dally, U. Kapasi, S. Rixner, P. Mattson, and B. Mowery. "Polygon Rendering on a Stream Architecture." In *Proceedings of SIGGRAPH/Eurographics Workshop on Graphics Hardware*, pp. 23–32. New York: ACM Press, 2000.

[72] B. Pailthorpe, N. Bordes, W.P. Bleha, S. Reinsch, and J. Moreland. "High-Resolution Display with Uniform Illumination." *Proceedings Asia Display IDW*, pp. 1295–1298.

[73] K. Perrine and D. Jones. "Parallel Graphics and Interactivity with the Scaleable Graphics Engine." In *Proceedings of the 2001 ACM/IEEE Conference on Supercomputing*, p. 5. New York: ACM Press, 2001.

[74] C. S. Pinhanez, F. C. Kjeldsen, A. Levas, G. S. Pingali, M. E. Podlaseck, and P. B. Chou. "Ubiquitous Interactive Graphics." Technical Report RC22495 (W0205-143), IBM Research, 2002.

[75] C. Pinhanez, M. Podlaseck, R. Kjeldsen, A. Levas, G. Pingali, and N. Sukaviriya. "Ubiquitous Interactive Displays in a Retail Environment." In *Proceedings of SIGGRAPH Sketches*. New York: ACM Press, 2003.

[76] C. Pinhanez. "The Everywhere Displays Projector: A Device to Create Ubiquitous Graphical Interfaces." In *Proceedings of Ubiquitous Computing, Lecture Notes in Computer Science 2201*, pp. 315–331. Berlin: Springer-Verlag, 2001.

[77] C. Poynton. *A Techical Introduction to Digital Video.* New York: John Wiley and Sons, 1996.

[78] A. Raij, G. Gill, A. Majumder, H. Towles, and H. Fuchs. "PixelFlex2: A Comprehensive, Automatic, Casually-Aligned Multi-Projector Display." In *IEEE International Workshop on Projector-Camera Systems*. Washington, D.C.: IEEE Computer Society, 2003.

[79] R. Raskar, G. Welch, M. Cutts, A. Lake, L. Stesin, and H. Fuchs. "The Office of the Future: A Unified Approach to Image-Based Modeling and Spatially Immersive Display." In *Proceedings of SIGGRAPH 98, Computer Graphics Proceedings, Annual Conference Series*, pp. 168–176. Reading, MA: Addison Wesley, 1998.

[80] R. Raskar, M. S. Brown, R. Yang, W. Chen, H. Towles, B. Seales, and H. Fuchs. "Multi-Projector Displays Using Camera-Based Registration." In *Proceedings of IEEE Visualization*, pp. 161–168. Washington, D.C.: IEEE Computer Society, 1999.

[81] R. Raskar, J. van Baar, and J. Chai. "A Low-Cost Projector Mosaic with Fast Registration." In *Asian Conference on Computer Vision*, pp. 23–25, 2002.

[82] R. Raskar, J. van Baar, P. Beardsley, T. Willwacher, S. Rao, and C. Forlines. "iLamps: Geometrically Aware and Self-Configuring Projectors." In *ACM SIGGRAPH 2006 Courses*. New York: ACM Press, 2003.

[83] R. Raskar. "Immersive Planar Displays using Roughly Aligned Projectors." In *Proceedings of IEEE Virtual Reality 2000*, pp. 109–116. Washington, D.C.: IEEE Computer Society, 2000.

[84] R. Raskar. "Projector-Based Three Dimensional Graphics." Technical Report TR02-046, University of North Carolina at Chapel Hill, 2001.

[85] L. Renambot, A. Rao, R. Singh, B. Jeong, N. Krishnaprasad, V. Vishwanath, V. Chandrasekhar, N. Schwarz, A. Spale, C. Zhang, G. Goldman, J. Leigh, and A. Johnson. "SAGE: The Scalable Adaptive Graphics Environment." In *Proceedings of WACE*, 2004.

[86] M. Román, C. K. Hess, R. Cerqueira, A. Ranganathan, R. H. Campbell, and K. Nahrstedt. "A Middleware Infrastructure to Enable Active Spaces." *IEEE Pervasive Computing* 4:1 (2002), 74–83.

[87] R. Samanta, J. Zheng, T. Funkhouser, K. Li, and J. P. Singh. "Load Balancing for Multi-Projector Rendering Systems." In *SIGGRAPH/Eurographics Workshop on Graphics Hardware*, pp. 107–116. New York: ACM Press, 1999.

[88] H. Y. Shum and R. Szeliski. "Construction of Panoramic Mosaics with Global and Local Alignment." *International Journal of Computer Vision* 36:2 (2000), 101–130.

[89] P. Steurer and M. B. Srivastava. "System Design of Smart Table." In *Proceedings of the First IEEE International Conference on Pervasive Computing and Communications*, p. 473. Washington, D.C.: IEEE Computer Society, 2003.

[90] M. C. Stone. "Color and Brightness Appearance Issues in Tiled Displays." *IEEE Computer Graphics and Applications* 21:5 (2001), 58–66.

[91] M. C. Stone. "Color Balancing Experimental Projection Displays." In *9th IS&T/SID Color Imaging Conference*, pp. 342–347. Springfield, VA: Society for Imaging Science and Technology, 2001.

[92] M. C. Stone. *A Field Guide to Digital Color*. Natick, MA: A K Peters, 2003.

[93] N. A. Streitz, J. Geißler, T. Holmer, S. Konomi, C. Müller-Tomfelde, W. Reischl, P. Rexroth, P. Seitz, and R. Steinmetz. "i-LAND: An Interactive Landscape for Creativity and Innovation." In *Proceedings of the SIGCHI Conference on Human Factors in Computing Systems*, pp. 120–127. New York: ACM Press, 1999.

[94] E. H. Stupp and M. S. Brennesholtz. *Projection Displays*. New York: John Wiley and Sons Ltd., 1999.

[95] R. Sukthankar, R. Stockton, and M. Mullin. "Automatic Keystone Correction for Camera-Assisted Presentation Interfaces." In *Proceedings of International Conference on Multimedia Interfaces (ICMI), Lecture Notes in Computer Science 1948*, p. 607. Berlin: Springer-Verlag, 2000.

[96] R. Sukthankar, R. Stockton, and M. Mullin. "Smarter Presentations: Exploiting Homography in Camera-Projector Systems." In *Proceedings of International Conference on Computer Vision (ICCV)*, pp. 247–253. Washington, D.C.: IEEE Computer Society, 2001.

[97] R. Surati. "Scalable Self-Calibration Display Technology for Seamless Large-Scale Displays." Ph.D. thesis, Department of Computer Science, Massachusetts Institute of Technology, 1999.

[98] J. Underkoffler, B. Ullmer, and H. Ishii. "Emancipated Pixels: Real-World Graphics in the Luminous Room." In *Proceedings of SIGGRAPH 99, Computer Graphics Proceedings, Annual Conference Series*, pp. 385–392. Reading, MA: Addison Wesley Longman, 1999.

[99] Russell L. De Valois and Karen K. De Valois. *Spatial Vision*. Oxford University Press, 1990.

[100] *VR Juggler Suite*. Available online (http://www.vrjuggler.org/), 2000.

[101] G. Wallace, H. Chen, and K. Li. "Color Gamut Matching for Tiled Display Walls." In *Immersive Projection Technology Workshop*, 2003.

[102] R. Yang and G. Welch. "Automatic Projector Display Surface Estimation Using Every-Day Imagery." In *9th International Conference in Central Europe on Computer Graphics, Visualization and Computer Vision*, 2001.

[103] R. Yang, D. Gotz, J. Hensley, H. Towles, and M. S. Brown. "PixelFlex: A Reconfigurable Multi-Projector Display System." In *Proceedings of IEEE Visualization*, pp. 167–174. Washington, D.C.: IEEE Computer Society, 2001.

Index

Printed and bound by CPI Group (UK) Ltd, Croydon, CR0 4YY

23/10/2024

01777698-0001